S0-BIU-794

DELMAR PUBLISHERS • ALBANY, NEW YORK 12205
A DIVISION OF LITTON EDUCATIONAL PUBLISHING, INC.

Preface

In 1895, it was discovered that a combination of oxygen and acetylene produced a flame with a higher temperature than any flame yet known. At about the same time methods of producing oxygen and calcium carbide were invented. These two developments opened the field of oxyacetylene welding. Before this time most metal joining was done with rivets.

Since the invention of the oxyacetylene welding process there have been many variations of the process and new methods of producing the heat to fuse metal have been developed. However, the oxyacetylene flame continues to be one of the most important sources of heat for joining and cutting metals. The ability to use oxyacetylene equipment efficiently and safely is considered a requirement for entry into many trades. WELDING PROCEDURES: OXYACETYLENE is designed to help the student acquire these abilities.

Each unit of the text is preceded by clearly stated, behavioral objectives which describe what the student will be able to do when the unit is completed. All of the subsequent material in the unit is designed to fulfill these objectives.

A concise discussion of any pertinent related information follows the unit objectives. This part of the unit contains such things as new terms, safety information, and a description of problems that might be encountered.

At the heart of each unit is an opportunity for the student to learn while performing a job with oxyacetylene equipment. These jobs are designed to enhance one another, each building on the knowledge and skills gained through those preceding it. The jobs include a complete list of all necessary equipment and material; a procedure for performing an oxyacetylene operation; key points, to make the operation more meaningful and to aid the student in the successful completion of the job; and, in most cases, a destructive test of the completed joint.

At the conclusion of each unit there are review questions. By answering these questions, the student can review some highlights of the unit and be sure all of the material has been learned, without requiring special testing by the instructor.

In addition to the unit features of the textbook, each of the five sections is concluded with a comphrensive review. The comprehensive review consists of two or more procedures to be followed. In addition, the instructor's guide contains questions about the material presented throughout the section. These procedures and questions provide a logical means by which the instructor can test the student's knowledge and ability.

Frank Schell has been a journeyman welder for twenty-four years. He is currently Curriculum Development Coordinator and Professor of Welding at the College of Southern Idaho. He is a member of the American Vocational Association and the Idaho Vocational Association. He is the author of several pieces of instructional material for welding students. WELDING PROCEDURES: OXYACETYLENE is a product of the author's background in welding, both as an educator and as a journeyman.

Safety Rules

Oxygen Handling

1. Important. Use no oil or grease around oxygen. When mixed with oil or grease, oxygen can cause a violent explosion.
2. Do not use pipe fitting compounds on oxygen connections.
3. Only hoses which are made for welding should be used.
4. Do not force connections which do not fit.
5. If gas cylinders are not clearly marked as to contents, do not use them.
6. Do not use oxygen under high pressure without an oxygen regulator.
7. Be sure that the oxygen cylinder is securely fastened so it cannot fall.
8. Be sure the pressure adjusting screw is turned out (to the left) before the cylinder is opened.
9. Stand to one side when opening a cylinder.

Acetylene Handling

1. Do not use pipe fitting compounds on acetylene equipment.
2. Always use an acetylene pressure reducing regulator.
3. Never force connections which do not fit.
4. Be sure the acetylene cylinder is securely fastened so it cannot fall.
5. Do not release acetylene into the atmosphere when welding is being done in the area.
6. Leave the bottle key on the acetylene cylinder.

General Rules

1. Always use a striker to light a torch.
2. Never lay a lighted torch down.
3. Before lighting the torch be sure it is not pointing at another person.
4. Before lighting the torch be sure that the flame will not come in contact with inflammable material.
5. Make it a habit to hold your hand close to a piece of metal to see if it is hot, before picking it up.
6. Never operate equipment without instruction on its use.
7. Wear safety glasses or grinding shields when using a power grinder.

8. Be aware of the condition of the hoses on the torch. If a leak develops it should be reported immediately, and the hoses taken out of service.

9. Report all burns.

10. Never work with defective equipment.

11. Hammers, chisels and punches wear out. Do not use them if they are defective.

12. Never use any kind of fire around oxygen and acetylene cylinders. Oxygen supports combustion and acetylene will burn or explode.

13. Bronze rod contains zinc. Galvanizing on some steel contains zinc. When zinc is heated it gives off a toxic vapor. Do not breathe these fumes. Be sure the ventilation is adequate.

14. Avoid breathing the fumes when welding painted objects.

15. Always wear goggles when cutting or welding.

16. Always wear protective clothing when cutting or welding.

17. Know the location of fire extinguishers, and how to use them.

18. Some factors which contribute to accidents:
    Poor lighting
    Improper electric outlets
    Poor maintenance of equipment
    Poor ventilation
    Poor housekeeping
    Machines without guards
    Poor arrangement of equipment
    Horseplay
    Slippery, dirty floors
    Damaged equipment
    Inadequate instruction.

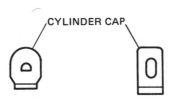

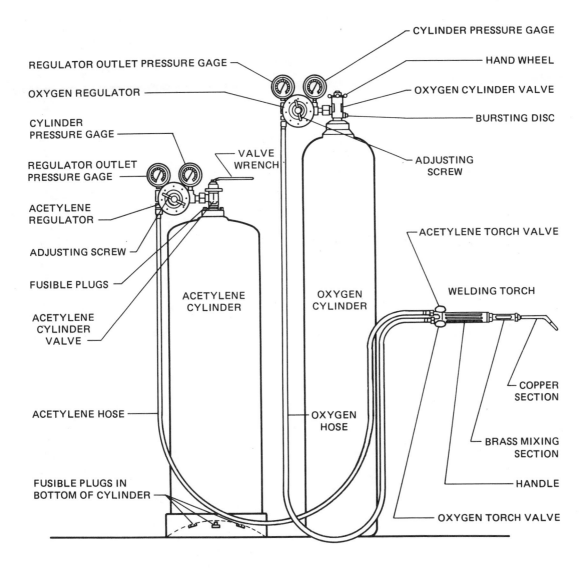

Contents

SECTION 1:
Welds in the Flat Position

Unit 1   Setting Up Oxyacetylene Welding Equipment

OBJECTIVES

After completing this unit the student will:

- Properly set up oxyacetylene welding equipment.
- Observe specific safety precautions when handling oxyacetylene equipment.
- Be able to list the components of an oxyacetylene welding outfit and describe their functions.

The following is a list of the most important equipment for a standard welding outfit:

- Oxygen
- Acetylene
- Oxygen Regulator
- Acetylene Regulator
- Welding Hoses
- Torch (sometimes called a blowpipe)
- Welding tips
- Goggles
- Hammer
- Pliers
- Striker
- Tip cleaners
- Table with steel or firebrick top
- Twelve-inch adjustable wrench (or manufacturer's cylinder wrenches)
- Protective clothing

Fig. 1-1 Oxygen Cylinder

Oxyacetylene Processes

Oxyacetylene welding is based on the principle that, when acetylene gas is burned in the proper proportions, with oxygen gas, a flame is produced,

1

which is hot enough to melt and fuse metals. This proportion is approximately 1 part of acetylene to 2½ parts of oxygen. Oxyacetylene flame cutting uses much of the same equipment, but the principle is different. In oxyacetylene cutting a stream of oxygen is directed against a piece of *ferrous* metal (metal which contains iron is called ferrous), which has been heated to a red heat. This causes the metal to burn.

Oxygen Cylinder

The oxygen cylinder, figure 1-1, is usually green or yellow in color, so that it can be identified. It is made from a single plate of high grade steel, which has been heat treated to develop toughness and strength. When fully charged, the oxygen *bottle*, as it is some-times called, contains 244 cubic feet of oxygen at a pressure of 2,200 pounds per square inch at 70° Fahrenheit. This oxygen is 99.99% pure and is colorless, odorless and taste-less. Oxygen by itself will not burn, but it does support combustion.

> CAUTION: Because of the extremely high pressure at which oxygen is stored in the cylinder, several precautions must be observed at all times.

- All cylinders must have Interstate Commerce Commission markings, indicating the dates of bottle pressure tests.
- Cylinders must be stored so they cannot be knocked down.
- They should not be stored in an area where extreme temperature changes occur.
- Oxygen cylinders must not be stored near grease, oil or electrical connections. Bringing oxygen into contact with oil or grease may cause a violet explosion.
- They must never be moved without the cylinder cap in place, on top of the cylinder.
- Cylinders which are defective in any way should be taken out of service and re-ported to the supplier.

Valve Protection Cap

The valve protection cap, or *bottle cap*, screws onto the cylinder and completely covers the valve, figure 1-2. It protects the valve from damage when the cylinder is being moved or if it is accidentally knocked over.

Oxygen Cylinder Valve

The oxygen cylinder valve, figure 1-3, is attached to the top of the oxygen cylin-der. It is used to turn the flow of oxygen on or off, as needed. These valves are *double seated*. This means that when the valve is completely closed the flow of oxygen from the cylinder is shut off, and when the valve is opened all the way the valve seats and pre-vents leakage of oxygen around the valve stem. The valve should be opened complete-ly when the cylinder is in use.

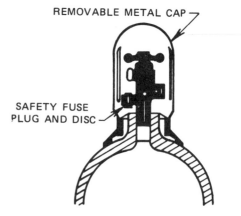

Fig. 1-2 Oxygen Cylinder Cap and Safety Plug

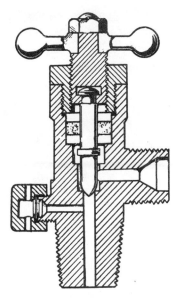

Fig. 1-3 Oxygen Cylinder Valve

A *safety fuse plug* and *disc* are installed in the oxygen cylinder valve. As the temperature of the oxygen in the cylinder increases, the pressure also increases. If the pressure of the gas in the bottle becomes too great, the safety plug and disc will release the pressure.

Acetylene Cylinder

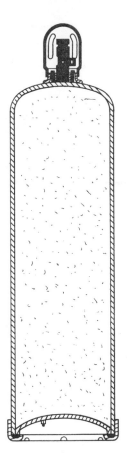

Fig. 1-4 Acetylene Cylinder

The pressure in acetylene cylinders is not as high as that in oxygen cylinders. For this reason, acetylene cylinders are rolled to the size needed and welded at the seams, figure 1-4.

Acetylene cylinders are filled with a porous material, such as fuller's earth or balsa wood. A liquid chemical (acetone) is poured into the bottle and is absorbed by the porous material. Acetone absorbs acetylene gas.

Acetylene gas, which is made by mixing water and calcium carbide (a gray, rock-like substance), has a strong disagreeable odor, resembling garlic. It is highly inflammable and, in combination with oxygen, produces the hottest flame known (5,800° - 6,300° Fahrenheit). When there is not enough oxygen present it burns with a smoky, yellow flame.

> CAUTION: Because acetylene gas is highly inflammable and explosive, certain safety precautions must be observed.
>
> - Cylinders must be tested and certified by the Interstate Commerce Commission.
> - Cylinders which leak or are defective in any way should be taken out of service and reported to the supplier.
> - Free acetylene gas (that which is not absorbed in acetone) must not be stored at pressures above 15 pounds per square inch. Above this pressure acetylene becomes very unstable and may explode.
> - If large numbers of acetylene cylinders are stored close to oxygen cylinders a fire resistant wall must be built between the two types of cylinders.

- Cylinders should never be used in any position but the upright position. Liquid acetone can run into the gages and hoses if the cylinder valve is opened while the cylinder is lying on its side.
- Never store acetylene cylinders where excessive heat may contact them.

Acetylene Cylinder Valves

The acetylene cylinder valve, figure 1-5, is attached to the top of the acetylene cylinder. It is used to turn the flow of acetylene on or off as needed. Acetylene cylinder valves are not double seated, because they do not have to withstand the high pressure that oxygen cylinders do.

These valves are of two types. One type has a handwheel, resembling that on the oxygen cylinder valve. The other has a square stem, without the wheel, and is turned on and off with a special wrench, called a *key*.

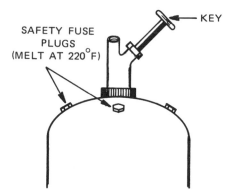

Fig. 1-5 Acetylene Bottle Key and Fuse Plugs

CAUTION: The acetylene cylinder valve should never be opened more than 1½ turns. In this way, it can be turned off quickly in case of fire. For the same reason, the key should always be left on the valve.

Acetylene cylinders have plugs installed in them for safety. These plugs are made of a metal which melts at a low temperature. Any excessive heat, which would cause the gas in the cylinder to reach higher pressure, melts the plugs. This allows the acetylene to escape and prevents an explosion.

Oxygen Regulators

Full oxygen cylinder pressure is 2,200 pounds per square inch. It is impossible to weld with this much pressure, so a regulator is installed on the cylinder, figure 1-6. This regulator allows the welder to set the pressure at reduced amounts. It has a safety device, which vents the pressure if it exceeds safe limits.

Regulators are equipped with two gages, figure 1-7. One indicates the cylinder pressure, while the other indicates the working or torch pressure. Oxygen gauges are generally built to withstand 3,000 pounds per square inch of pressure. When temperature variations cause the gas in the cylinder to expand, and increase the pressure, the gage indicates this rise in pressure.

The regulator is equipped with a nut, which screws onto the cylinder valve. The threads are conventional, right-hand threads. To install the regulator, tighten the nut with the wrench supplied by the manufacturer or with an adjustable wrench.

Acetylene Regulators

Acetylene regulators are similar to oxygen regulators, with two exceptions. All acetylene fittings have left-hand threads. This is important to remember, as the fittings may be damaged by attempting to turn them the wrong way. The reason for the left-hand threads on acetylene fittings, is to prevent them from being accidentally installed on oxygen equipment.

The second way in which they are different from oxygen regulators is that acetylene gauges have lower numbers than oxygen gauges. As a general rule, the acetylene cylinder pressure gauge registers to 500 pounds per square inch. The acetylene working pressure gauge registers to a maximum of 15 pounds per square inch. Also, the numbers and graduations on the dial of oxygen gauges are normally green, while on acetylene gases they are normally red.

Note: Gauges are delicate mechanisms and through mishandling they may not register correctly. However, the regulator will hold the correct pressure, even if the gage does not indicate it correctly.

Torch

The welding torch (or *blowpipe*) has separate inlets for oxygen and acetylene. It allows the gases to mix in the correct proportion for welding. The mixing is controlled by two valves on the handle, each of which may be opened and closed to regulate the flow of gases for the welding flame.

The most commonly used torch is the equal pressure (or medium pressure) type, figure 1-8. In this type approximately equal amounts of oxygen and acetylene are used for welding.

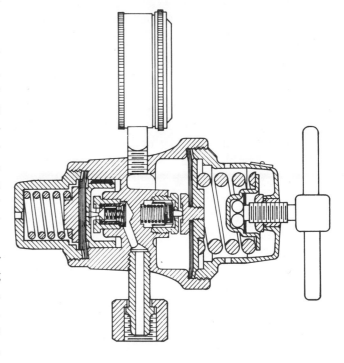

Fig. 1-6 Oxygen Regulator

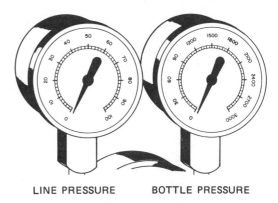

LINE PRESSURE   BOTTLE PRESSURE

Fig. 1-7 Oxygen Gauges

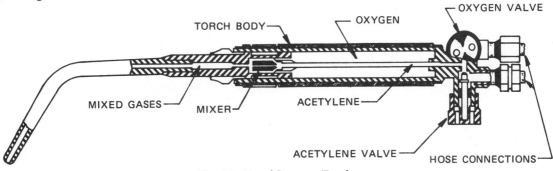

Fig. 1-8 Equal Pressure Torch

As with all other oxyacetylene equipment the torch has left-hand threads on the acetylene connection and right-hand threads on the oxygen connection.

Welding Tip

Different sized tips must be attached to the torch. These tips have a mixing chamber and *orifices* (holes) to supply different flame sizes. The tip concentrates the gases coming from the torch so that the flame can be directed toward the weld to be made.

Tips come with different size orifices, to vary the size of the flame. When a tip with larger holes is used more gas can escape. This provides a larger flame for welding on heavier metal, where more heat is required. Tips should be kept clean and in good working order.

Fig. 1-9 Bottle Cart

Bottle Cart

A bottle cart, figure 1-9, is a two-wheeled cart equipped with a chain for fastening the cylinders securely. When oxyacetylene welding equipment is installed on the bottle cart it makes a portable welding station which can be wheeled to the job.

Hoses

Special hoses are used for oxyacetylene welding equipment. They are made to withstand the high welding pressures. Welding hoses are supplied in 3/16", 1/4", 3/8" and 1/2" sizes. Select the correct size for the equipment to be used. They may be furnished as a double hose, connected by a rubber web and molded together. If single hoses are used, tape them together at 18-inch intervals to keep the unit solid.

Oxygen hoses are green in color and have right-hand threads. Acetylene hoses are red in color and have left-hand threads. Also, acetylene hose fittings have grooves cut into the nuts, to indicate a left-hand threaded connection.

SUMMARY

- Oxygen does not burn, but supports combustion.
- Oxygen in contact with oil or grease may cause a violent explosion.
- Oxygen is colorless, odorless and tasteless.
- Oxygen is bottled under extremely high pressures; the cylinders must be handled carefully.
- Acetylene is highly inflammable.
- Acetylene has a strong, disagreeable odor, resembling garlic.
- Acetylene is very unstable over 15 pounds per square inch.
- The oxyacetylene flame is the hottest flame known.
- Oxygen valves are double seated and should be either fully opened or fully closed.

- Acetylene valves should not be opened more than 1½ turns.
- Oxygen regulators register high pressures.
- Acetylene regulators are made to register lower pressures.
- Oxygen equipment is identified by green markings and has right-hand threaded connections.
- Acetylene equipment is identified by red color and has left-hand threaded connections, with grooves cut in the nuts.

JOB 1: SETTING UP OXYACETYLENE WELDING EQUIPMENT

Equipment:

Oxygen bottle
Acetylene bottle
Oxygen regulator
Acetylene regulator
One set of welding hoses
Torch
Bottle cart
Acetylene bottle key (if required)
12″ adjustable wrench or manufacturer's supplied wrenches

PROCEDURE	KEY POINTS
1. Obtain full oxygen and acetylene cylinders and install them in the bottle cart, or fasten them securely in an upright position.	1. Fasten the cylinders securely.
2. Remove the cap from the oxygen cylinder. Store the cap on the cart.	
3. Open the cylinder valve slightly. Allow a small amount of oxygen to blow through the valve. Close the cylinder valve.	3. Do not stand in front of the valve. Make sure no one else is standing in front of the valve. This is called *cracking* the valve.
4. Install the oxygen regulator, using a suitable wrench.	4. Oxygen cylinders and regulators have right-hand threads. Tighten the nut snugly, but do not apply too much force or the threads may be stripped.
5. Remove the cap from the acetylene cylinder. Store the cap.	
6. Open the valve on the acetylene cylinder slightly, using the cylinder key. Close valve.	6. **CAUTION:** This gas is inflammable. No open fire!

PROCEDURE	KEY POINTS
7. Install the acetylene regulator.	7. Left-hand thread.
8. Install the green hose on the oxygen gauge; install the red hose on the acetylene gauge.	8. Tighten fittings snugly. Over tightening will strip the threads.
9. Open the oxygen cylinder valve slowly until a small amount registers on the gauge, then open it completely. Turn the adjustment screw on the regulator to the right, until a small amount of pressure shows on the low pressure gauge. This will blow the hose clean. Turn the adjustment screw to the left and release the pressure.	9. Sudden release of oxygen pressure may damage the gauges. Open valve very slowly until a slight pressure registers. **CAUTION:** Do not stand in front of the gauge face. Pressure may blow face outward.
10. Open the acetylene valve slowly until a small amount registers on the gauge. Then open it 1 1/2 turns. Turn the adjustment screw on the regulator to the right, until a small amount of pressure shows on the low pressure gauge. This will blow the hose clean. Turn the adjustment screw to the left and release the pressure.	10. Acetylene gas is inflammable. When releasing it into the room, be sure there is no open fire present. Do not open the acetylene valve more than 1 1/2 turns.
11. Install the blowpipe (torch) on the open ends of the hoses.	11. Acetylene fittings have left-hand threads.
12. Be sure the torch valves are closed. Adjust the regulators so that 10 pounds of pressure shows on both the acetylene and the oxygen gauges.	
13. Check all connections with soapsuds. If bubbles appear, a leak is indicated and the connections must be tightened.	13. **CAUTION:** Do not use soap with an oil base. Oil and oxygen may cause a violent explosion. Use no oil.

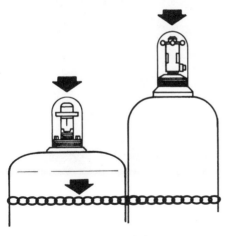

1. TWO BOTTLES SAFETY CHAINED AND CAPS
   IN PLACE

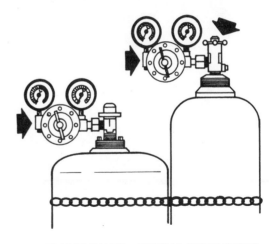

2. CRACK VALVES — INSTALL REGULATORS

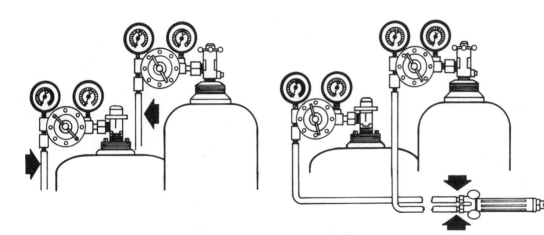

3. INSTALL HOSES.

4. INSTALL TORCH.

5. OPEN BOTTLE VALVES.

6. SET ADJUSTING SCREWS FOR LINE PRESSURE.

Fig. 1-10

Summary: Job 1

- Fasten the bottles securely in an upright position.
- Crack the valves to clean any dirt from them.
- Install the regulators. (Acetylene has left-hand threads)
- Install the hoses.
- Install the torch.
- Set the regulators for 10 pounds per square inch, working pressure.
- Check for leaks with soapsuds.
- Reverse the procedure to disassemble the equipment.

REVIEW QUESTIONS

1. List the steps to be followed in setting up a welding outfit.
2. What are the main differences between the valve on an oxygen cylinder and the valve on an acetylene cylinder.
3. What is acetylene made from?
4. What is the maximum safe pressure of free acetylene?
5. Why are cylinder valves cracked before installing the regulators?
6. Why is "Use No Oil" so strongly stressed?
7. Why should the oxygen cylinder valve be opened very slowly after the regulator is attached?
8. What is the purpose of a regulator?
9. What is the difference between oxygen and acetylene fittings?
10. Is oxygen inflammable?

Unit 2 Lighting the Oxyacetylene Torch

OBJECTIVES

The welding student will:

- Light the oxyacetylene torch following proper safety precautions.
- Be able to identify oxidizing, carburizing, and neutral flames.
- Adjust the torch for each of the three types of flame.
- Discuss the effect of each of the three types of flame on the metal.

Flames

Three basic flames can be made by adjusting the valves on the welding torch.

- The *carburizing flame.*
- The *neutral flame.*
- The *oxidizing flame.*

Carburizing Flame

A *carburizing flame*, figure 2-1, is the result of too much acetylene gas in the flame. This flame may be recognized by a long streamer of green colored gas which burns around the inner cone of the flame. This is called an acetylene feather. A carburizing flame is used to make the outside of metal hard, but is not good for a weld. The addition of the extra acetylene to the melted weld adds carbon to the metal and makes a hard, brittle weld. When this flame is used on melted parent metal, it causes the puddle to turn dark red, and gives it a boiling action.

Neutral Flame

The *neutral flame,* figure 2-2, is the welding flame. This flame can be recognized by a sharp inner cone and the absence of an acetylene feather. It is made up of 2 1/2 parts of oxygen and 1 part of acetylene. One part of the oxygen in the flame and one part of the acetylene come from the bottles. The other 1 1/2 parts of oxygen are picked up from the air around

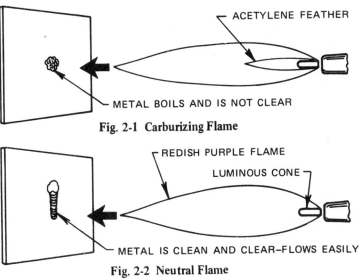

ACETYLENE FEATHER

METAL BOILS AND IS NOT CLEAR

Fig. 2-1 Carburizing Flame

REDISH PURPLE FLAME

LUMINOUS CONE

METAL IS CLEAN AND CLEAR–FLOWS EASILY

Fig. 2-2 Neutral Flame

the welding tip. A neutral flame does not add anything to or subtract anything from the *parent metal* (the metal being welded). The acetylene torch is adjusted for a neutral flame, for most welding jobs that require the metal to be melted and mixed together.

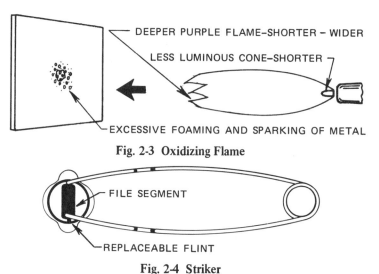

Fig. 2-3 Oxidizing Flame

Fig. 2-4 Striker

Oxidizing Flame

An *oxidizing flame,* figure 2-3, is the result of having too much oxygen in the gas mixture. This flame can be identified by a shorter inner cone and a whistling sound. It causes the molten metal to boil and spark. The additional oxygen in the flame causes the metal to burn, resulting in a brittle weld. A slightly oxidizing flame may be used for brazing, but it is not used for fusion welding.

Striker

The striker, figure 2-4, produces a spark by dragging a piece of flint across a file. A striker must always be used to light the oxyacetylene torch. The use of matches creates a hazard and may result in personal injury. The flints in most strikers may be replaced when the original one is worn out.

Goggles

Goggles are to be worn when welding. They are made in many shapes and sizes, to suit the individual welder. Welding goggles have dark lenses which filter out ultraviolent and infrared rays. For oxyacetylene welding they should be equipped with #4 or #5 filter lenses, either blue or brown. A clear glass cover lens fits over the filter lens, to protect the more expensive filter lenses from hot weld spatter.

Ultraviolet rays are given off by the welding flame. These are the same invisible rays which make it dangerous for the human eye to look directly at the sun. These rays can cause severe burns and possible blindness.

Infrared rays can cause a burn which looks like sunburn. The skin must be covered by clothing and the eyes protected by dark lenses to avoid the dangers of burns from infrared rays while welding.

JOB 2: LIGHTING THE OXYACETYLENE TORCH

Equipment:

    Oxyacetylene outfit as assembled in Unit 1
    Welding tips
    Goggles
    Gloves
    Striker

PROCEDURE	KEY POINTS
1. Set up the welding equipment, following the procedures outlined in JOB 1.	1. Review KEY POINTS of JOB 1.
2. Install the tip on the torch body.	2. Tighten the tip retaining nut hand tight only.
3. Open the oxygen cylinder valve slightly until pressure registers on the high pressure gauge, figure 2-5, then open the valve fully.	3. Do not stand in front of the gauges when opening the cylinder valve.

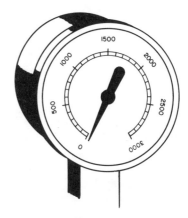

Fig. 2-5 High Pressure Gauge (Oxygen)

4. Turn the adjusting screw on the regulator to the right until pressure registers on the low pressure gauge, figure 2-6.	4. Set the oxygen pressure at about 5 psi.

Fig. 2-6 Low Pressure Gauge (Oxygen).

5. Open the oxygen needle valve on the torch handle and readjust the regulator until about 5 pounds registers, with the needle valve open. Close the needle valve.	5. Always adjust the pressure with torch valve open. When it is closed, pressure may register higher on the gauge, but the welding pressure will be correct with the valves open.

6. Open the acetylene cylinder valve slightly until pressure registers on the high pressure gauge, figure 2-7, then open it 1/2 turn.

6. Never open the acetylene cylinder valve over 1/2 turn. Leave the key on the valve.

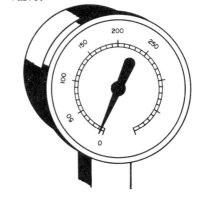

Fig. 2-7 High Pressure Gauge (Acetylene)

7. Turn the adjusting screw to the right until pressure registers on the low pressure gauge, figure 2-8.

Fig. 2-8 Low Pressure Gauge (Acetylene)
(Note Calibration Stops at 15 lbs.)

8. Open the acetylene needle valve on the torch handle and readjust the screw on the regulator until about 5 pounds registers, with the needle valve open. Close the needle valve on the torch.

8. Set at about 5 psi.

9. Open the acetylene needle valve on the torch about 1/2 turn. Hold the striker in the left hand (if right-handed), the torch in right hand, and strike a spark in front of the escaping gas.

9. Be sure the torch is not pointed toward any inflammable material or people. Wear gloves and goggles.

10. Open the needle valve on the torch until the flame jumps away from tip about 1/8".

11. Open the oxygen needle valve on the torch slowly, adding oxygen to the burning acetylene.

12. As oxygen is added to the acetylene, observe the luminous cone at the tip, and the long greenish-color envelope around it. The green envelope is the excess acetylene of the carburizing flame.

13. Continue to add oxygen by opening the oxygen needle valve until the feather of acetylene just disappears. The inner cone will now appear soft and luminous. This is a neutral flame.

14. If more oxygen is added now, the flame becomes pointed and white in color. In addition, it makes a sharp whistling sound. This is an oxidizing flame.

15. Practice adjusting the torch to carburizing, neutral, and oxidizing flames.

16. Shut off the acetylene needle valve on the torch; shut off the oxygen needle valve on the torch; shut off both cylinder valves completely; open the needle valves on the torch handle and drain the hoses. (Watch the gauges until they register 0.)

17. Close the needle valves on the torch, release the pressure adjusting screws on the regulators by turning the handles to the left; coil the hoses and hang them up on the hose holder.

10. The flame will appear turbulent, but will not smoke.

12. A carburizing flame is used for hardening steel. It is generally used with the acetylene feather about 3 times as long as the inner cone.

13. The torch should make a soft, even blowing sound.

14. An oxidizing flame is harmful to the weld. It is never used for welding.

16. Shut off the acetylene first, so the escaping oxygen will blow any soot or impurities from tip.

17. Do not hang the hoses on the gauges. The weight may damage the gauges. Make sure no fires are burning around the work area.

Summary: Job 2

- Set up the welding equipment.
- Install the tip on the torch.
- Open the oxygen cylinder valve slowly.

- Turn the adjusting screw on the regulator until the gauge indicates 5 psi.
- Open the oxygen needle valve on the torch.
- Readjust the gauge pressure to 5 psi with the valve open.
- Open the acetylene cylinder valve.
- Turn the adjusting screw on the regulator until the gauge indicates 5 psi.
- Open the acetylene needle valve on the torch.
- Readjust the pressure to 5 psi with the valve open.
- Close the valve on the torch.
- Open the acetylene valve on the torch and light the flame.
- Continue to open the acetylene needle valve until the flame jumps slightly away from the tip.
- Open the oxygen needle valve on the torch and adjust the flame to neutral.
- Reverse the procedures for shutting off the equipment.
- Check for fires.

REVIEW QUESTIONS

1. What is a neutral flame?
2. What is a carburizing flame?
3. What is an oxidizing flame?
4. Why is the pressure on the line gauges adjusted with the needle valves open?
5. When shutting off the torch, which needle valve should be closed first? Why?
6. Why should the operator stand behind, or to one side of, the oxygen cylinder when opening the cylinder valve?
7. What is the maximum working pressure of acetylene gas?

Unit 3   Running a Bead with Filler Position Rod, Flat

OBJECTIVES

The student will:

- Be able to define penetration and its importance.
- Be able to describe filler rod and its use.
- Be able to describe the flat position for welding.
- Run a bead in the flat position using filler rod.

Penetration

An oxacetylene weld should always have 100% penetration. This means that the weld appears on the bottom of the parent metal as well as on the top. Poor penetration causes the metal to break in the weld. However, too much penetration causes the molten metal to drip through and hang down below the parent metal. This condition is known as "icicles". Excessive drop-through of molten metal causes oxidation, as the molten metal takes oxygen out of the atmosphere into the weld. This oxidation causes a brittle weld which is easily broken.

Mild Steel

Steel which contains 0.30% carbon, or less, is called mild steel. It is the most commonly used steel for construction purposes. It is also called low-carbon steel and black iron. Most of the steel used for construction purposes has a black appearance, caused by hot rolling at the steel mills, resulting in the formation of oxide on the metal.

Forehand Welding

Right-handed welders usually hold the torch in the right hand and weld from right to left, adding the filler rod at the front of the flame. This procedure is called *forehand* welding. Left-handed operators weld from left to right, but the rod is still added at the front of the flame, since the torch is held in the left hand and the filler rod in the right.

Flat Weld Position

A weld made on the topside of the parent metal and within 30 degrees of horizontal is called a *flat weld*. Flat welding is the most desirable position for welding, since the welder can control penetration and bead appearance easily. In many welding shops, machines called *positioners* are used to hold the work so that it can easily be turned into the flat position.

Filler Rod

Filler rod is commonly called welding rod. It is added to the molten puddle to build up the cross section of weld where the penetration has forced the molten metal below the surface of the parent metal. To insure good penetration, filler rod should be added only after the puddle has been formed. The weld should have a cross section thicker than the original parent metal, so that strength is added at the point of the weld. The word *convex* means a rounded-up bead. Filler rods most commonly used are 1/16″, 3/32″ and 1/8″ diameters.

Sheet Metal

Metal which has been rolled in the steel mill to a thickness of 1/8″ or less is commonly called sheet metal. This is the metal most commonly used with the oxyacetylene welding process, since any thickness over 1/8″ is more easily joined by electric arc welding. Sheet metal is designated by gauges. A few of the most common gauges are listed below:

12-gauge . . . .1120″	approx. 1/8″	thickness
14-gauge . . . .0821″	approx. 5/64″	thickness
16-gauge . . . .0635″	approx. 1/16″	thickness
18-gauge . . . .0508″	approx. 3/64″	thickness

Neutral Flame

A neutral flame is used for welding. On an equal pressure torch, the gauges should be set at approximately the same pressure, if the welding is not to be done in a confined area. Welding in a restricted area, such as a corner of the metal, uses up the atmospheric oxygen rapidly and changes the character of the flame from neutral to carburizing. More oxygen pressure may be needed for such a welding condition. In this case the oxygen pressure must be higher than the acetylene pressure, until the welding in the restricted area is completed. If the oxygen pressure has been increased, it must be lowered after the welding in the corner is finished.

JOB 3: RUNNING A BEAD WITH FILLER ROD

Equipment and Material:

Standard Oxyacetylene Welding Equipment.
16-gauge mild steel sheet metal
1/16″ mild steel filler rod.

PROCEDURE	KEY POINTS
1. Place the sheet metal sample on a firebrick.	
2. Light the torch and adjust to a neutral flame.	
3. Wear goggles to protect the eyes. Hold the torch in the right hand, if right-handed, and a piece of filler rod in left hand.	3. Grasp rod as you would a pencil, with 6″ to 8″ extending below the left hand.

PROCEDURE	KEY POINTS

4. Hold the flame 1/16" to 1/8" above the parent metal at the right edge of the sample until a molten puddle is formed. Dip the end of the rod into the puddle, allowing a small amount of the rod to melt off and fuse with the base metal, figure 3-1.

4. Do not add rod until a molten puddle is formed. Keep the end of the rod in the outer end of flame envelope to protect it from oxidation. The tip should be at an angle of 45° to 60° with the parent metal, figure 3-2.

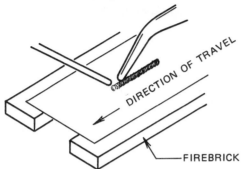

Fig. 3-1 Welding a Bead on Sheet Metal

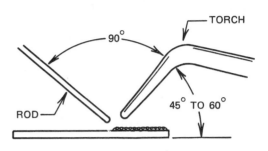

Fig. 3-2 Torch Angle for Welding

5. Weld across the sample, from right to left, adding filler rod to the molten parent metal.

5. This is called forehand welding.

6. Melt enough rod into the puddle to build up the bead evenly in height and width. Alternately raise the rod as the base metal melts and dip it into the front of the puddle when penetration is completed.

6. Get penetration before adding the filler rod. Filler metal melted onto the sheet metal is not a weld, unless it is fused into the parent metal. A slight circular motion of the flame will make a better weld, figure 3-3.

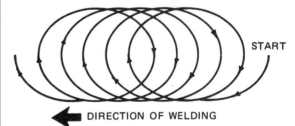

DIRECTION OF WELDING

Fig. 3-3 Circular Torch Movement

7. When good penetration is obtained the underside of the metal will look as though a bead has been run on it. Addition of the rod to the molten puddle is the key to good welding. Practice this job until the technique is mastered.

7. The edges of the weld should be feathered smoothly into the parent metal. The rod should not be piled up. The bead should look like fish scales overlapping each other.

Summary: Job 3

- Weld with a neutral flame.

- Weld from right to left (left to right if left-handed).

- The torch should be at an angle of 45° to 60° with the parent metal.

- Circular motion of the flame is used to control the heat and bead appearance.

- If the metal becomes too hot the torch may be flashed off the weld. Momentarily direct the flame away from the puddle, but keep the puddle fluid.

- When good penetration has been developed, the underside of the sheet metal will look like a bead has been run on it.

REVIEW QUESTIONS

1. What is parent metal?

2. Why is it necessary to get penetration before the rod is added to the weld?

3. Why is the addition of filler rod necessary for a strong weld?

4. What type of flame is used for welding?

5. What percentage of penetration is necessary for the best weld?

Unit 4   Butt Weld on Mild Steel Sheet Metal, Flat Position

OBJECTIVES

The student will be able to:

- List and define several terms applying to butt welding.
- Discuss the effects of expansion and contraction.
- Compensate for expansion and contraction in butt welding.
- Make a butt weld which withstands a bend test.
- Perform a bend test on a butt weld.

Gapping

One of the greatest concerns the welder has in butt welding steel sheet metal is how expansion and contraction are dealt with. When steel is heated, as in welding, it *expands,* or increases in size. As it cools it *contracts,* or decreases in size.

When butt welding sheet metal it is a good practice to gap the metal. At the side where the weld is to begin leave about a 1/16-inch gap between the two pieces of parent metal. At the side where the weld is to end leave about a 1/8-inch gap. (These figures apply to a piece of 16-gage sheet metal 6-inches long.) The extra gap at the finishing end of the weld allows for the additional heat which is absorbed by the metal as the weld moves along.

Tack the pieces securely, so the gap will remain for welding. Tack welds are small completely fused welds made at several places along the line of the weld to hold the pieces in place.

Attempting to weld sheet metal without gapping the butt weld will produce a scissors effect, where one piece of sheet metal will be warped over the top of the other.

Keyhole

At the start of the weld, as the puddle is formed, the edges ahead of the weld should be melted away a small amount. This creates a place for the penetration to work through the gap. This keyhole should remain small. As it is carried along in front of the puddle, it insures 100% penetration.

Testing

A good weld will bend 180° over the bead, without breaking and will stand as much pull as the parent metal when tested for tensile strength. Most of the tests used in this text are *destructive tests.* They will bend the welded jobs out of shape, so the metal cannot be used again without considerable preparation.

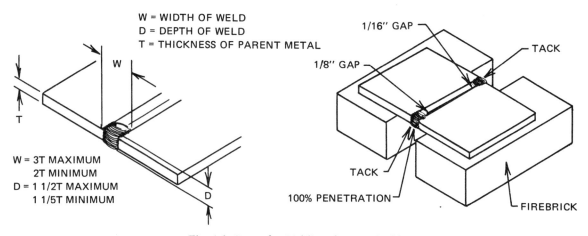

Fig. 4-1 Setup for Welding the Butt Weld

JOB 4: BUTT WELD IN MILD STEEL SHEET METAL, FLAT POSITION

Equipment and Material:

> Standard Oxyacetylene Welding Equipment
> 16-gauge mild steel samples, 2″ wide by 6″ long
> 1/16″ or 3/32″ mild steel filler rod

PROCEDURE	KEY POINTS
1. Tack weld two pieces of 16-gauge metal as shown in figure 4-1.	1. Leave a gap between the pieces.
2. Beginning at right side, form a puddle and add a little rod. Progress across the joint, alternately melting the puddle and adding rod at the leading edge of the puddle.	2. Keep a keyhole ahead of the puddle at all times so that penetration is achieved before the rod is added to the puddle. 100% fusion and penetration are necessary for a strong weld.
3. Get complete fusion of the metal to the bottom of the gap, and add filler rod so that the weld will be above the top of the parent metal.	3. The weld should always be thicker than the parent metal.
4. Practice this exercise until the butt weld is mastered.	
5. Test the weld. Place the welded sample in a vise with the weld at the top of the vise jaws. Bend the metal against the bead a full 180° No cracks or tears should appear.	5. Both good appearance and strength are necessary. Be sure the sample is cold before testing. Hot metal bends easier, but has less strength.

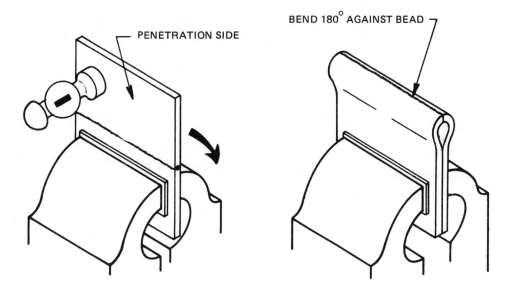

Fig. 4-2 Testing Butt Weld in Mild Steel Sheet Metal

REVIEW QUESTIONS

1. Why is it necessary to gap a butt weld?
2. How high should the bead be built up on this type of weld?
3. What percentage of penetration should be achieved?
4. What is the keyhole?
5. What type of flame is used for this weld?

Unit 5 Fillet Weld on Mild Steel Sheet Metal, Flat Position

OBJECTIVES

The student will be able to:

- Explain undercutting and how to prevent it.
- Discuss the consequences of a lack of oxygen in a confined area and tell how to compensate for this condition.
- Make a fillet weld capable of withstanding a bend test.
- Perform a bend test on a fillet weld.

Fillet Weld

A *fillet weld* is a weld made on two pieces of metal which are joined in any way other than in a flat plane. A fillet is a reinforcement and a weld made in an inside corner is called a fillet weld. Fillet welds are sometimes called T-welds, when one piece forms a 90° angle with the other.

The type of welded joint most often used is the fillet weld. Generally, the fillet is the most difficult type of joint to weld successfully. Penetration must be made completely through the corner to insure that the joint will have full strength. Sometimes welds are made on both sides of the upright piece, but in many cases it is not possible to reach both sides, so the weld made from one side must be strong enough to hold the pieces together.

Undercutting

Gravity exerts a force on fluid metal, so there is a tendency for the metal to drop away from the vertical piece on a T-weld. The metal will be made thinner by the amount that drops away. This thinned section is called an *undercut.*

All fillet welds are subject to undercutting. Great care must be taken to add enough filler metal to completely build up the bead and the torch must be directed into the weld in such a way that excessive melting does not take place in the upper part of the fillet joint. Undercutting, figure 5-2, reduces the thickness of the metal, making it subject to breakage under stress.

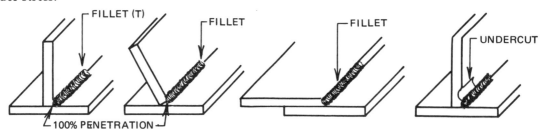

Fig. 5-1   Fig. 5-2

Popping the Torch

Because welding in confined areas, such as corners, exhausts the available oxygen in the atmosphere the torch tip may become overheated. Frequently the flame goes out momentarily from lack of oxygen. As it reignites, from the hot metal, a popping action results which blows the molten metal out of the weld. This condition may be corrected by (1) using a little more oxygen pressure on the torch, or (2) by changing to the next size larger tip. If more oxygen pressure is used, care must be taken so the flame does not become oxidizing and burn the weld.

JOB 5: FILLET WELD IN MILD STEEL SHEET METAL (T-WELD)

Equipment and Material

Standard oxyacetylene welding equipment
16-gauge mild steel sheet metal samples, 2″ x 6″
1/16″ or 3/32″ mild steel filler rod

PROCEDURE	KEY POINTS
1. Tack weld two pieces of metal as shown in figure 5-3. 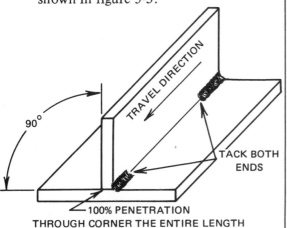 Fig. 5-3	
2. Beginning at the right side, play the flame on the flat piece until it is red; then with a slight circular motion, gradually heat both pieces until fusion is accomplished.	2. Since the weld is being made in a corner, more heat will be required. Use one size larger tip if necessary Do not direct the flame completely into corner. The burning gases will overheat the tip.
3. Move at a regular rate of speed across the joint, melting the two pieces together, then adding rod.	3. The rod should be added slightly above the center of the puddle to avoid undercutting.

TRAVEL DIRECTION
90°
TACK BOTH ENDS
100% PENETRATION THROUGH CORNER THE ENTIRE LENGTH

PROCEDURE	KEY POINTS
4. Practice this weld until penetration and bead appearance are mastered.	4. Be sure to fill the puddle completely. Fuse the edges of weld into both pieces of parent metal. Get complete fusion of parent metals when welding the corner before the rod is added. If the rod is piled up there is no way to get back into the corner to melt the parent metal.
5. Test the weld. Place the welded sample in a vise and bend both pieces of metal against the weld, figure 5-4. No cracks, tears or bead separation should appear.	

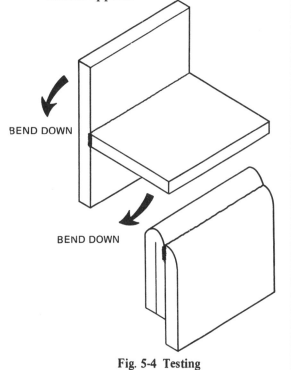

BEND DOWN

BEND DOWN

Fig. 5-4 Testing

REVIEW QUESTIONS

1. Why does this type of weld require more heat than a butt weld?

2. Why is there danger of overheating the tip in this exercise?

3. What is an undercut?

4. Why should undercutting be avoided?

5. What is a fillet weld?

6. What type of flame is used for this weld?

Unit 6 Lap Weld on Mild Steel Sheet Metal, Flat Position

OBJECTIVES

The student will be able to:

- Discuss burn-away and how it is prevented.

- Tell why lap welds are avoided where possible.

- Make a lap weld capable of withstanding a bend test.

- Perform a bend test on a lap welded joint.

A *lap weld* is a weld made with one plate of metal overlapping the other.

The lap joint is not considered to be a desirable method of construction. Even when the two pieces are clamped together tightly, there is often danger of moisture collecting in the space between them. This moisture promotes oxidation (rusting). Corrosion can also occur between the two pieces.

Burn-Away and Torch Angle

Because lap welds are made on the edge of one piece of metal and on the flat side of the other, the top piece absorbs heat faster than the bottom piece. This causes faster melt-

ing and burn-away of the edge of the top piece. Burn-away can be corrected by (1) pointing the flame onto the bottom piece, (2) keeping the filler rod between the flame and the edge of the top piece. The filler rod should be added from the top into the molten puddle formed on the bottom piece.

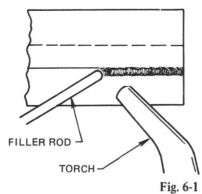

FILLER ROD

TORCH

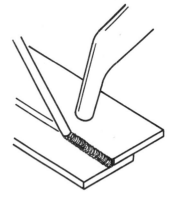

Fig. 6-1

JOB 6: LAP WELD IN MILD STEEL SHEET METAL, FLAT POSITION

Equipment and Material

Standard oxyacetylene welding equipment
16-gauge mild steel samples, 2″ x 6″
1/16″ or 3/32″ mild steel filler rod

PROCEDURE	KEY POINTS
1. Lay one piece of sheet metal over the other, so that it overlaps slightly, figure 6-2.	1. The two pieces of metal must fit tightly. Gaps will cause unsightly welds and poor fusion.

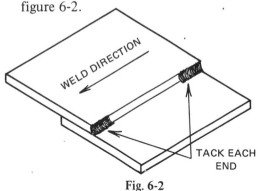

Fig. 6-2

2. Tack weld the two pieces together at each end.	
3. Start the weld on right side of the plate, melting the upper piece into the lower, and adding filler rod as necessary.	3. Play the flame on the lower piece. The upper edge will melt more rapidly, so the welding rod should be added at the upper edge. Build a full bead. A shallow bead will cause the weld to fail under stress.
4. Test the weld by bending both pieces of metal 180° over the face of the weld.	4. The weld will fail if it has not been built up enough, or if penetration is shallow in the lower plate.

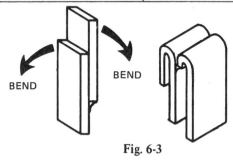

Fig. 6-3

REVIEW QUESTIONS

1. Why is the lap weld considered to be a poor joint design?

2. Why should two pieces of metal in a lap weld be clamped tightly together?

3. Where should the flame be directed in this type of weld?

4. Why is a full bead important on this type of joint?

5. What type of flame is used for this weld?

Unit 7   60° Fillet Weld on Mild Steel Sheet Metal, Flat Position

OBJECTIVES

The student will be able to:

- Discuss bridging and how it is prevented.
- Make a 60° fillet weld with 100% penetration and capable of withstanding a bend test.
- Bend test a 60° fillet welded sample.

A 60° weld is a fillet weld made in a corner where two pieces of metal join at a 60° angle.

This is a difficult weld to make. The lack of space between the two pieces restricts the torch motion and reduces combustion as the oxygen is burned away in the confined area. Extreme care must be taken to melt the parent metal in the corner completely before the filler rod is added.

Bridging

If the filler rod is added before penetration is achieved in the corner, it will bridge across the gap, rather than melting into it. A bridged weld will not stand a bend test, since the angled parent metal is not fused into the flat plate. It is important when welding in confined areas to have the molten puddle achieve complete fusion before the filler rod is added. Filler rod should be added in small quantities so the puddle is not cooled off by the filler rod.

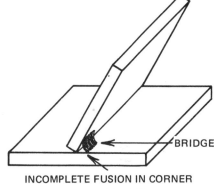

INCOMPLETE FUSION IN CORNER

Fig. 7-1

JOB 7: 60° WELD IN MILD STEEL SHEET METAL, FLAT POSITION

Equipment and Materials:

Standard oxyacetylene welding equipment
2 samples 16-gauge mild steel, 2" x 6"
1/16" or 3/32" mild steel filler rod

PROCEDURE

1. Tack weld one piece of sheet metal in the center of another, at an angle of 60°, figure 7-2.

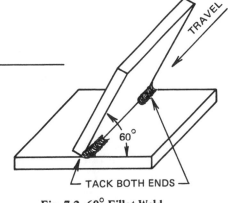

TACK BOTH ENDS

Fig. 7-2 60° Fillet Weld

PROCEDURE	KEY POINTS
2. Beginning at the right edge, make a smooth, continuous weld from one end to the other.	2. Add filler rod only after the puddle has been established. Use a tip one size larger than would ordinarily be required. The confined area requires more heat.
3. Keep the bead the same width.	3. Slow, careful concentration is required to make a good weld at this angle.
4. Test the weld, as in figure 7-3.	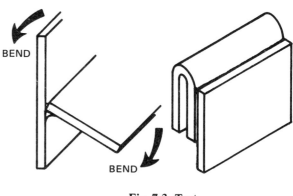

Fig. 7-3 Test

REVIEW QUESTIONS

1. Draw a sketch of a 60° fillet weld between two pieces of sheet metal. Show the angle between the two pieces.

2. What does the term bridging mean in the welding field?

3. What should be done if the metal does not melt in the corner of a 60° fillet weld?

SECTION 1: WELDS IN THE FLAT POSITION, COMPREHENSIVE REVIEW

A.   RUNNING A BEAD IN THE FLAT POSITION

Equipment and Materials:

> Standard oxyacetylene welding equipment
> 1 piece 16-gauge sheet metal, 2″ x 6″
> 1/16″ or 3/32″ mild steel filler rod

PROCEDURE

1. Place a piece of 16-gage metal in the flat position.

2. Using a neutral flame and 1/16″ filler rod, make a single pass bead across the 6″ length of the sheet metal.

3. Cool the sample and brush the weld with a wire brush.

4. Have the sample inspected by the instructor.

B.   BUTT WELDING IN THE FLAT POSITION

Equipment and Materials:

> Standard oxyacetylene welding equipment
> 2 pieces 16-gauge sheet metal, 2″ x 6″
> 1/16″ or 3/32″ mild steel filler rod

PROCEDURE

1. Tack weld 2 pieces of 16-gage sheet metal in position for a butt joint.

2. Using a neutral flame and 1/16″ filler rod, weld the two pieces together with a butt weld in the flat position.

3. Cool the sample and brush with a wire brush.

4. Have the sample inspected by the instructor for bead appearance and penetration.

5. Test the weld as shown in the drawing.

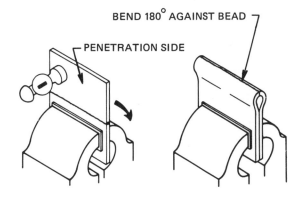

BEND 180° AGAINST BEAD

PENETRATION SIDE

SECTION 2:
Welds in Horizontal and Vertical Positions

Unit 8   Bead on Mild Steel Sheet Metal, Horizontal Position

OBJECTIVES

The student will be able to:

- Describe cold lap and its cause.
- Discuss the use of torch angle and filler rod manipulation in controlling cold lap.
- Run a horizontal bead with good penetration and free of cold lap.

A *horizonal weld* is a weld which is done in a horizontal line and against an approximately vertical surface, figure 8-1.

Welding in the horizontal position is more difficult than welding in the flat position, because the force of gravity tends to cause the molten metal to run down onto the plate. This run-down can be controlled by using the correct torch angle and feeding the rod correctly.

Torch Angle

For horizontal welding, the tip and flame should point slightly upward, with the tip positioned at an angle of 45° to 60° with the work. The filler rod should be fed from the top of the puddle.

Cold Lap

When the molten metal runs down onto the cold metal and it is not mixed with the parent metal, a cold lap results, figure 8-2. This is a problem in horizontal welding, because

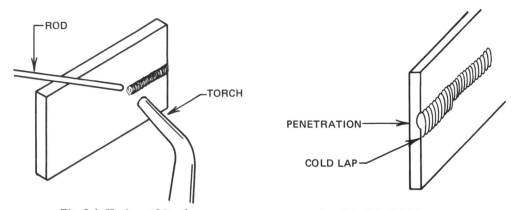

Fig. 8-1 Horizontal Bead   Fig. 8-2 Cold Lap

gravity tends to pull the hot metal downward. The edge of the bead rolls onto the parent metal, without fusion, resulting in a poor weld, with little or no strength.

JOB 8: RUNNING A HORIZONTAL BEAD IN MILD STEEL SHEET METAL

Equipment and Materials:

> Standard oxyacetylene welding equipment
> 2 pieces, 16-gauge mild steel, 2" x 6"
> 1/16" or 3/32" mild steel filler rod

PROCEDURE	KEY POINTS
1. Position a piece of steel as shown in figure 8-3.	1. The metal may be tacked in a T-position on another plate, to hold it upright. 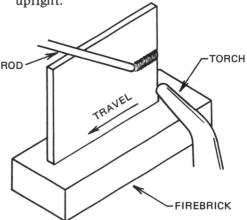 **Fig. 8-3 Position of Work for Running a Horizontal Bead.**
2. Beginning at the right side, weld a bead across the 6" length, being careful to get a smooth appearance and 100% penetration.	2. Special attention must be paid to the direction and angle of the torch. The filler rod must be added above the center of the puddle.
3. Examine the back of the metal to check the penetration of the bead.	

REVIEW QUESTIONS

1. What is cold lap?

2. What percent of penetration is required on a horizontal bead?

3. What type of flame is used to weld the horizontal bead?

Unit 9   Butt Weld on Mild Steel Sheet Metal, Horizontal Position

OBJECTIVES

The student will be able to:

- Manipulate the torch and filler rod to control cold lap.

- Make a butt weld in the horizontal position with 100% penetration.

A horizontal butt weld is similar to a butt weld in the flat position. The only difference is that the pieces to be welded are in a vertical position, with the joint running in a horizontal line.

Practice is necessary to master the art of horizontal butt welding. The welder must use the correct tips and pressures and handle the torch and filler rod skillfully. The horizontal butt weld is a common application of oxyacetylene welding. The welding student should concentrate on perfecting this type of weld.

Torch Angle and Cold Lap

Horizontal butt welding involves running a horizontal bead, so cold lap is controlled in the same manner. The torch is aimed upward, at an angle of 45° to 60° with the work, and the filler rod is fed from the top.

Crown

Weld beads should be built up above the surface of the parent metal to insure strength. This buildup is called the *crown* of the weld.

JOB 9: HORIZONTAL BUTT WELD IN MILD STEEL SHEET METAL

Equipment and Materials:

Standard oxyacetylene welding equipment
2 pieces, 16-gauge mild steel, 2" x 6"
1/16" or 3/32" mild steel filler rod

PROCEDURE	KEY POINTS
1. Tack weld two pieces of sheet metal for a butt weld.	1. The metal should be gapped approximately 1/16" at the right side and 1/8" at the left.

34

2. Place the material in a horizontal position, figure 9-1.

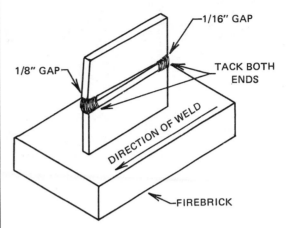

Fig. 9-1 Horizontal Butt Weld

3. Make a butt weld across the 6" length of the two pieces.

3. Be careful that the molten metal does not roll to the bottom of the bead. Weld slowly and carefully and add the filler rod from the top of the puddle.

4. The weld should appear uniform and smooth, with 100% penetration.

5. Test the weld according to figure 9-2.

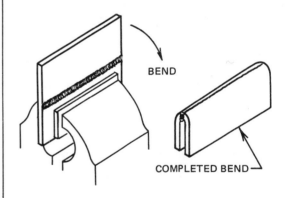

Fig. 9-2 Test

REVIEW QUESTIONS

1. What is the crown of a weld?

2. When making a horizontal butt weld, how much gap should be left between the two pieces of sheet metal?
   a. Right-hand side _____ ".
   b. Left-hand side _____ ".

3. When testing a butt weld, the sheet metal should be bent over the _____ side of the welded sample.

4. What is the proper torch angle for horizontal butt welding?

Unit 10   Bead on Mild Steel Sheet Metal, Vertical Position

OBJECTIVES

The student will be able to:

- Describe the vertical position for welding.
- Discuss methods to overcome the force of gravity in vertical welding.
- Run a bead in the vertical position with 100% penetration.

A vertical weld is any weld done in a vertical line. Vertical beads are usually run from the bottom to the top of the work.

Welding in the vertical position presents the problem of overcoming the force of gravity. Torch position and tip angle are used to overcome the tendency of the molten puddle to flow downward. The flame, fueled by gas under pressure, provides some support for the molten puddle. Feeding the filler rod from the top of the weld also helps, by providing a ledge of slightly hardened, cooled metal under the flame. As this ledge is formed the next layer of the weld is built upon it, figure 10-1.

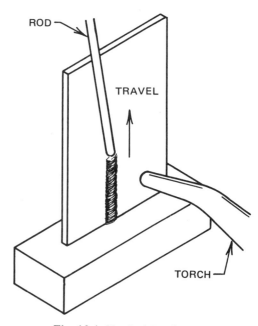

Fig. 10-1 Vertical Bead

JOB 10: VERTICAL BEAD ON MILD STEEL SHEET METAL

Equipment and Materials:

Standard oxyacetylene welding equipment
16-gauge sheet metal sample, 2" x 6"
1/16" or 3/32" mild steel filler rod

PROCEDURE	KEY POINTS
1. Place a piece of 16-gage sheet metal in a vertical position, so that the 6" length is perpendicular to the bench.	

PROCEDURE	KEY POINTS
2. Beginning at the bottom of the sheet, weld upward, adding rod as necessary.	2. To overcome the effects of gravity, which pulls the molten puddle down the sheet, add the rod from the side or top of the puddle. Keep the upward progress slow and even and do not add too much rod. The bead should be uniform in width, smoothly feathered in at the sides, and should have 100% penetration.

REVIEW QUESTIONS

1. Describe the position of the work in vertical welding.

2. How much penetration is needed on a vertical weld?

3. What is the purpose of the ledge formed in vertical welding?

4. Describe the action of the torch which helps to hold the puddle of molten metal from flowing downward.

5. What direction should the filler rod be added from when welding a vertical bead?

6. Is it possible for vertical beads to be welded so they are as strong as flat beads?

Unit 11 Butt Weld on Mild Steel Sheet Metal, Vertical Position

OBJECTIVES

The student will be able to:

- Define flash off and use it effectively.
- Restart a bead in the vertical position insuring uniform appearance of the finished bead.
- Make a vertical butt weld capable of withstanding a bend test.

A vertical butt weld is similar to a flat or horizontal butt weld, except that the weld is made in a vertical line.

A major difficulty when butt welding in the vertical position is the tendency of the edges to melt away and leave holes ahead of the torch. Careful application of the heat and the angle of the torch contribute to making a successful weld. Vertical butt welding is a common practice in sheet metal welding and must be mastered by the welding student.

Sheet metal should always be gapped and tack welded to prevent warpage. Welding in the vertical position requires concentration of the flame into a small area, so that the heat travelling up the joint does not become too intense and melt holes ahead of the puddle. The metal should be held rigidly in the vertical position by a clamp or holding a device called a *jig*.

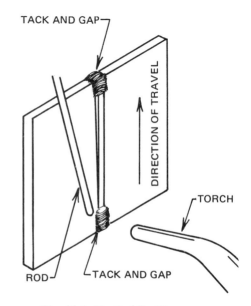

Fig. 11-1 Vertical Position

The filler rod should be melted into the puddle from above the center of the weld. This helps maintain the ledge, on which the weld progresses. Also, the torch tip must be directed upward into the weld, in the same manner as for running a vertical bead. This will help to prevent the molten metal from flowing down.

Flash Off

Frequent flash off from the weld prevents hole burning. However, when the flame is flashed off, care should be taken to insure that the puddle does not completely solidify. If the puddle does solidify, the flame should be concentrated on the bead slightly below the puddle. The bead should be remelted at this point and the welding started only after a new puddle has been formed. The weld should then progress upward into the place where the previous welding stopped. Using this method insures complete penetration where the welding stopped and eliminates cold laps, which might occur at starting and stopping points.

JOB 11: BUTT WELD IN MILD STEEL SHEET METAL, VERTICAL POSITION

Equipment and Materials:

Standard oxyacetylene welding equipment
2 pieces, 16-gauge sheet metal, 2" x 6"
1/16" or 3/32" mild steel filler rod

PROCEDURE	KEY POINTS
1. Tack weld two pieces of material together, leaving a gap, as required for butt welding.	1. The gap should be 1/16" at the bottom and 1/8" at the top.
2. Place the material in a jig, so that the joint is perpendicular to the ground.	2. If no jig is available tack the work to a piece of scrap metal, so that it is held in a vertical position.
3. Make a continuous weld from bottom to top, being careful to maintain an even bead and 100% penetration.	3. Work in a comfortable position. The hoses may be draped over the welder's shoulders, to relieve the wrists of the weight of the hoses. Weld at the same rate of speed as for a flat butt weld. The heat travelling ahead may have a tendency to melt the metal faster.
4. Flash off the weld if there is a tendency to burn holes.	
5. Test the weld the same as a flat butt weld.	

REVIEW QUESTIONS

1. What is done to prevent warpage of sheet metal during butt welding?

2. What is the greatest difficulty encountered in butt welding in the vertical position?

3. From what direction should the filler rod be melted into the puddle of a vertical weld?

4. Describe how the torch flame is used to help support the molten puddle.

5. What is a jig?

6. How is a vertical weld restarted to prevent cold lap?

7. What should the percentage of penetration be on a vertical butt weld?

8. What may be done to prevent hole burning?

Unit 12 Lap Weld on Mild Steel Sheet Metal, Vertical Position

OBJECTIVES

The student will be able to:

- Manipulate the flame correctly to prevent burn away in vertical lap welding.

- Manipulate the filler rod to help control heat in vertical lap welding.

- Lap weld a joint in the vertical position which can withstand a bend test.

A vertical lap weld is a weld made between two pieces of metal, one overlapping the other with the joint running in a vertical line.

This is a difficult weld to make, but it can be mastered with practice. A finished vertical lap weld should have the same appearance as one made in the flat position. The bead should be smooth and the rippled, fish scale appearance uniform.

The raw edge of the overlapping plate has a tendency to burn away rapidly while the flat surface of the other plate is heating. The torch flame must be directed at the flat surface of the second plate. When the molten puddle is started it is carried along in the direction of travel.

The filler rod should be added from above the puddle and between the flame and the raw edge of the overlapping piece, figure 12-1. In this manner the filler rod will absorb some of the heat which might burn away the raw edge. The gas pressure will help support the molten puddle. Progress up the weld should be fairly rapid.

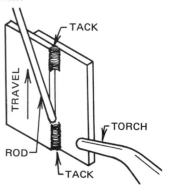

Fig. 12-1 Vertical Lap Weld Bend Test

JOB 12: LAP WELD IN MILD STEEL SHEET METAL, VERTICAL POSITION

Equipment and Materials:

Standard oxyacetylene welding equipment
2 pieces, 16-gauge sheet metal, 2" x 6"
1/16" or 3/32" mild steel filler rod

PROCEDURE	KEY POINTS
1. Position the metal samples with one piece lapped slightly over the other and tack weld.	

PROCEDURE	KEY POINTS
2. Weld the joint in the vertical position, from bottom to top.	2. Direct the torch at the surface of the second plate, melting the edge off the overlapping plate. Do not add too much rod; just enough to fill the bead completely.
3. Test the weld as in figure 12-2.	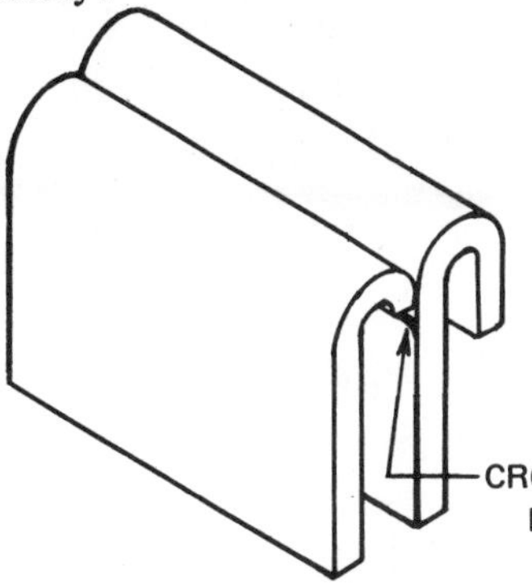 CROWN OF BEAD

Fig. 12-2 Vertical Lap Weld Bend Test

REVIEW QUESTIONS

1. How should the appearance of a finished vertical lap weld compare with the appearance of one made in the flat position?

2. What percentage of penetration is required for full strength in a lap weld?

3. To make a good vertical lap weld, which plate should the flame be directed toward?

4. What direction should the filler rod be added from when making a vertical lap weld?

5. What are the reasons that proper torch angle is important when making a vertical lap weld?

Unit 13   T-Weld on Mild Steel Sheet Metal, Vertical Position

OBJECTIVES

The student will be able to:

- Make a vertical T-weld with 100% penetration.
- Make a vertical T-weld without melting holes in either plate.
- Make a vertical T-weld capable of withstanding a bend test.
- Discuss the importance of preheating in making a vertical T-weld.

A *vertical T-weld* is a fillet weld made between two pieces which are at a 90° angle with each other, with the joint running in a vertical line, figure 13-1.

A vertical T-weld requires careful concentration. The corner, where the two pieces of metal join, must be completely penetrated. Areas in which the penetration is poor will cause the weld to fail when it is tested. To insure complete penetration, the heat should be directed into the corner, so that complete fusion of the two plates is accomplished before the filler rod is added.

Pointing the flame of the torch at one plate more than the other tends to burn holes in the metal. Flash off of the torch is necessary when there is danger of burning through the sheet metal, but 100% penetration is important for full-strength joints.

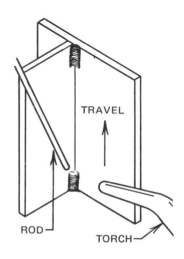

Fig. 13-1 Vertical T-Weld

As with other vertical welds, the torch should be tipped upward slightly and the filler rod added from above the puddle. This helps to support the molten puddle.

JOB 13: T-WELD IN MILD STEEL SHEET METAL, VERTICAL POSITION

Equipment and Materials

Standard oxyacetylene welding equipment
2 pieces, 16-gauge sheet metal, 2″ x 6″
1/16″ or 3/32″ mild steel filler rod

PROCEDURE

1. Tack weld one piece of sheet metal to the other piece at a 90° angle. Position the first piece approximately in the center of the second, with the joint running perpendicular to the welding table, figure 13-2.

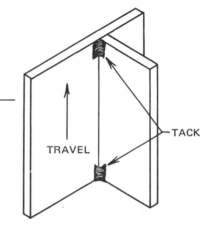

Fig. 13-2 Setup of T-Weld

PROCEDURE	KEY POINTS
2. Weld from the bottom to the top.	2. Hold the torch so the heat travels up the joint. Preheating helps penetration.
3. Concentrate on obtaining 100% penetration and smooth bead appearance.	
4. Test the weld as shown in figure 13-3.	

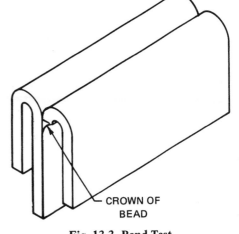

CROWN OF
BEAD

Fig. 13-3 Bend Test

REVIEW QUESTIONS

1. How much penetration of a T-weld is necessary for full strength?

2. A T-weld is also called a _____ weld.

3. What will happen if the flame is pointed at one plate of a T-weld more than the other?

4. In order to preheat the metal, the flame should be pointed _____ slightly when making a vertical T-weld.

5. Why should the metal be preheated when making a vertical T-weld?

6. What is the direction of travel on a vertical T-weld?

SECTION II: WELDS IN HORIZONTAL AND VERTICAL POSITIONS, COMPREHENSIVE REVIEW

A. HORIZONTAL BUTT WELD

Equipment and Materials

Standard oxyacetylene welding equipment
2 pieces, 16-gauge mild steel, 2″ x 6″
1/16″ or 3/32″ mild steel filler rod

PROCEDURE

1. Tack weld two pieces of sheet metal for a butt weld.

PROCEDURE

2. Position the material for a horizontal weld. See drawing of horizontal butt weld.

3. Make a butt weld across the 6" length of the two pieces.

4. Cool the weld and have it inspected by the instructor.

5. Test the weld according to the procedure given for testing a butt weld.

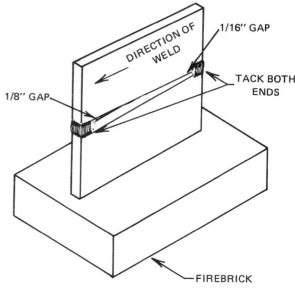

Setup for Horizontal Butt Weld

B. VERTICAL LAP WELD

Equipment and Materials

Standard oxyacetylene welding equipment
2 samples, 16-gauge mild steel, 2" x 6"
1/16" or 3/32" mild steel filler rod

PROCEDURE

1. Tack weld the samples at each end with one piece overlapping the other slightly as shown in the drawing of a vertical lap weld.

2. Position the material for a vertical weld.

3. Weld the joint from bottom to top.

4. Cool the weld and have it inspected by the instructor.

5. Test the weld according to the procedure given for testing a lap weld.

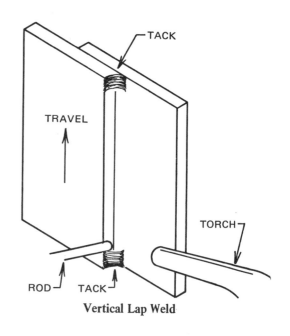

Vertical Lap Weld

SECTION 3:
Overhead Welds, Braze Welding, and Backhand Welding

Unit 14   Bead on Mild Steel Sheet Metal, Overhead Position

OBJECTIVES

The student will be able to:

- List the additional safety precautions necessary for overhead welding.
- Describe the overhead welding position.
- Run a bead in the overhead position with 100% penetration.

An *overhead weld* is one which is made on the underside of the joint and running in a horizontal line. Welds which are inclined 45° or less are considered to be in the overhead position, figure 14-1.

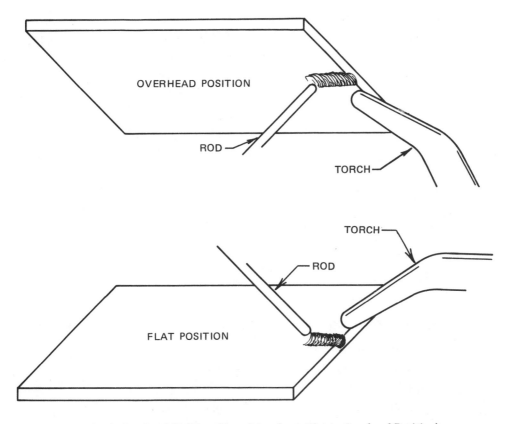

Fig. 14-1 Overhead Welding (Transition from Flat to Overhead Position)

CAUTION: There is a great danger of burns when welding in the overhead position. Sparks and hot metal fall on the welder frequently. Flame retardant clothing must always be worn.

Overhead welding is done with as much ease as flat welding. The awkward position of working with the arms above the head, with no support, makes it seem more difficult.

The weld should be made from right to left (for right-handed welders). The flame should be close to the metal and pointed slightly in the direction of travel. The filler rod should be added in small quantities.

The weight of the hoses may tire the welder. Draping the hoses over the shoulder will relieve this weight and the torch will handle more easily. A jig may be used to hold the material in position for overhead welding.

JOB 14: BEAD ON MILD STEEL SHEET METAL, OVERHEAD POSITION

Equipment and Materials
Standard oxyacetylene welding equipment
16-gauge mild steel sample, 2″ x 6″
1/16″ or 3/32″ mild steel filler rod

PROCEDURE	KEY POINTS
1. Use a jig to clamp a 16-gage sheet metal sample in the overhead position.	
2. Weld completely across the 6″ length with one bead, adding filler rod as necessary.	2. **CAUTION: Overhead welding may be dangerous. Sparks and molten metal fall from the weld. Wear protective clothing and stay alert.** The weight of the hoses may be carried on the welders shoulders. Use only enough rod to build up a slight crown.
3. Practice this until overhead welding is mastered.	

REVIEW QUESTIONS
1. Describe the burn hazards which may be encountered when welding in the overhead position.
2. Why does overhead welding seem more difficult than flat welding?
3. When overhead welding, the flame should be:
   a. Close and pointed in the direction of travel.
   b. Far away and pointed directly into the puddle.
   c. Close and pointed directly into the puddle.
   d. Far away and pointed in the direction of travel.

4. What percent of penetration is required when welding in the overhead position?

5. How much crown should an overhead weld have?

Unit 15   Butt Weld on Mild Steel Sheet Metal, Overhead Position

OBJECTIVES

The student will be able to:

- Handle the hoses in a manner that prevents the wrists from tiring.
- Manipulate the torch and filler rod in a manner that prevents holes from being melted through the parent metal.
- Make an overhead butt weld capable of withstanding a bend test.

An overhead butt weld, figure 15-1, is the same as a butt weld in the flat position, except that the weld is made on the underside of the parent metal. Overhead welds are made at an angle of not more than 45° from horizontal.

CAUTION: Sparks and hot metal may drop from the weld. Wear protective clothing.

The flame should be directed into the gap and the torch should be angled so the flame points in the direction of the weld. The filler rod should be added from the front of the puddle in small quantities. The bead should be small with about a 1/16" crown.

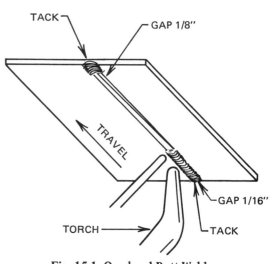

The following suggestions may help prevent holes from melting through the parent metal:

- Use the unmelted filler rod to cool the puddle.

- Flash the torch off the puddle momentarily.

- Use a smaller torch tip.

Fig. 15-1 Overhead Butt Weld

Using a smaller torch tip will also help control the amount of molten metal dripping from the weld. However, care must be used to insure that the tip is large enough to allow for 100% penetration.

JOB 15: BUTT WELD ON MILD STEEL SHEET METAL, OVERHEAD POSITION

Equipment and Materials

Standard oxyacetylene welding equipment
2 pieces, 16-gauge mild steel, 2" x 6"
1/16" or 3/32" mild steel filler rod

PROCEDURE	KEY POINTS
1. Prepare the material for a standard butt weld.	1. Gap the pieces the same as for a butt weld in the flat position. The tack welds should be solid.
2. Using any necessary jigs, clamp the material in the overhead position. The wider part of the gap should be on the left side (for right-handed welders).	
3. Weld the joint in one pass, adding filler rod as necessary.	3. Use only enough rod to build up a slight crown. The weight of the hoses may be carried on the welder's shoulders. Be sure penetration is 100%.
4. Test the weld as shown in figure 15-2.	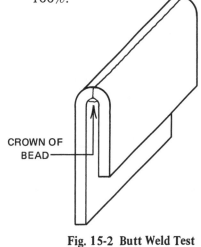 CROWN OF BEAD

Fig. 15-2 Butt Weld Test

REVIEW QUESTIONS

1. Why is welding in the overhead position dangerous?

2. Where should the flame be directed when butt welding in the overhead position?

3. From what direction should the filler rod be added to the puddle?

4. How much crown should be built up on an overhead butt weld?

5. What can be done to control burn through when making an overhead butt weld?

6. How many passes should be used to complete an overhead butt weld on 16-gage sheet metal?

7. How can the weight of the hoses be relieved from the wrists when welding in the overhead position?

Unit 16   Lap Weld on Mild Steel Sheet Metal, Overhead Position

OBJECTIVES

The student will be able to:

- Use oxyacetylene welding equipment with a smooth rhythm.
- Manipulate the flame and filler rod correctly to make an overhead lap weld.
- Make an overhead lap weld capable of withstanding a bend test.

An overhead lap weld is made on the underside of the metal and runs in a horizontal direction, figure 16-1. This is a difficult weld to make. It requires control of the flame and the puddle, and the filler rod must be added in a smooth rhythm.

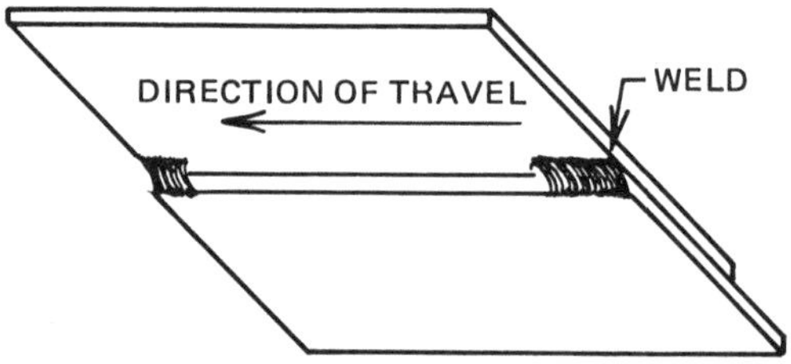

Fig. 16-1 Overhead Lap Weld
(Tack Both Ends)

Welding Rhythm

Welding rhythm is developed through practice of oxyacetylene welding. It is the ability to progress across a weld smoothly and uniformly. The puddle is melted and the filler rod is added in a rhythm which produces a smooth, rippled bead. Frequent pauses or the addition of rod in a ragged manner will spoil the appearance of a weld and also prevent uniform penetration of the joint.

The following is a list of hints for making a good overhead lap weld:

- Be careful to avoid burning away the edge of the lapped metal.
- Do not add filler rod too fast. Adding the filler rod too fast will cool the puddle and result in a bead with poor appearance. However, in order to withstand destructive testing, the bead must be full.
- The heat must be directed onto the flat surface of the second piece and the torch should always be pointed in the direction of travel.
- Keep the flame close to the metal.
- Drape the hoses over the shoulder to remove their weight from the wrist.

JOB 16: LAP WELD IN MILD STEEL SHEET METAL, OVERHEAD POSITION

Equipment and Materials

Standard oxyacetylene welding equipment
2 pieces, 16-gauge mild steel, 2″ x 6″
1/16″ or 3/32″ mild steel filler rod

PROCEDURE	KEY POINTS
1. Prepare the material for a standard lap weld.	1. Tack the samples at both ends.
2. Using any necessary jigs, clamp the material in the overhead position.	
3. Beginning at the right edge, weld the joint in one pass.	3. **CAUTION: Sparks and hot metal may drop from the weld. Wear protective clothing** Keep the flame pointed at the flat surface of the second plate, so the raw edge of the other plate melts slowly. Add only enough rod to fill the bead without cooling the puddle too much.
4. Test the weld as shown in figure 16-2.	 BEAD **Fig. 16-2 Overhead Lap Weld Test**

REVIEW QUESTIONS

1. What is meant by welding rhythm?

2. What might cause the appearance of an overhead lap weld to be poor?

3. Should an overhead lap weld have a full or shallow bead?

4. Where should the heat be directed when making an overhead lap weld?

5. How many passes should be used to complete an overload lap weld on 16-gage mild steel?

6. Describe the destructive test for an overhead lap weld.

Unit 17   Corner Welds on Mild Steel Sheet Metal

OBJECTIVES

The student will be able to:

- Weld a corner joint using no filler rod and obtaining 100% penetration.
- Down weld a joint with 100% penetration.
- Close small holes which may appear in a corner weld.
- Explain why down welding is not done on butt and lap joints.

Melt Welds

One of the most often used applications of oxyacetylene welding is the corner weld. Outside corners are frequently joined by *melt welding.* This refers to fusing two pieces without the use of filler rod. Smooth, strong beads, with excellent appearance can be made on sheet metal corners with little or no filler rod added.

When the pieces have been tacked, the flame is used to start a puddle. Then, with the flame pointing slightly in the direction of travel, the torch is moved across the joint, continuously melting the two edges, figure 17-1.

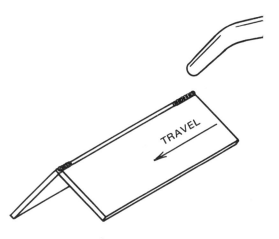

Fig. 17-1 Corner Weld-Flat Position

Care should be taken to keep the motion of the torch slow enough so that the melted metal can soak through the corner. This will insure 100% penetration.

If a hole develops in a weld, a drop of metal, melted from 1/16″ filler rod, may be added. This will close the hole, so the bead can be carried on across the joint.

Tip Size

Tip size is very important for corner welding. Usually, a small tip will give excellent results. Remember, however, that a puddle must be formed and the penetration must be 100%.

Position

Corner welds can be made in any of the four positions with little difficulty. In addition, corner welds are sometimes made in the vertical position from the top down. This is called *down welding.* Down welding is never done on any other type joint, since penetration is more difficult on the lap or butt weld, using this technique.

JOB 17: MELT WELDING CORNERS IN MILD STEEL SHEET METAL

Equipment and Materials

Standard oxyacetylene welding equipment
4 pieces, 16-gauge sheet metal, 2″ x 6″
1/16″ mild steel filler rod

PROCEDURE	KEY POINTS
1. Tack weld the sheet metal samples to make two assemblies, as shown in figure 17-2.	1. Tack the pieces together securely, using no filler rod. Melt the corners of the samples to form the tack welds.

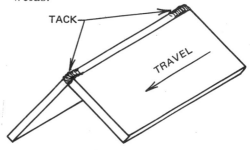

Fig. 17-2 Tack Welded Corner Joint

2. Weld one of the assemblies in the position shown in figure 17-2.	2. Using a neutral flame, start a puddle and work it slowly across the joint. Penetration should be 100%.
3. Down weld the other assembly, as shown in figure 17-3.	3. Weld from the top to the bottom of this joint.

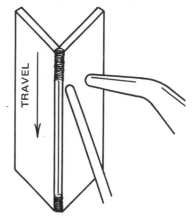

Fig. 17-3 Down Weld, Corner Joint

4. Do not use filler rod unless a hole appears which cannot be closed with parent metal.

PROCEDURE	KEY POINTS
5. Test both welds according to figure 17-4.	5. The welds should be smooth and uniform. 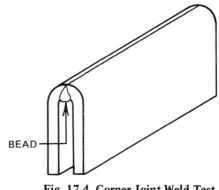 **Fig. 17-4 Corner Joint Weld Test**

REVIEW QUESTIONS

1. What is melt welding?
2. What is down welding?
3. For what reason is filler rod used in melt welding?
4. What type of flame is used to weld corner joints?
5. Why is down welding not recommended for welding butt and lap joints?

Unit 18   Braze Welding Sheet Metal

OBJECTIVES

The student will be able to:

- Describe the difference between brazing, braze welding, and fusion welding.
- Tell how and why flux is used in braze welding and brazing.
- Define capillary attraction.
- Make a bead with bronze filler metal.

Brazing

Brazing is a process in which a bond is produced between two pieces, which are in very close contact with each other, using a nonferrous filler metal. This filler metal has a lower melting point than that of the parent metal. As the filler metal melts it is drawn into the joint by *capillary attraction.*

Capillary Attraction

Capillary attraction is the property that allows a liquid to be drawn into a close fitting joint. This is different from the *cohesive* action that causes filler metal to adhere to a surface in braze welding. In order for this capillary attraction to take place it is necessary for pieces to be brazed to be in close contact with one another.

Braze Welding

In *braze welding* capillary attraction does not occur and the joint is designed the same as for fusion welding. The metal parts are joined by a molecular union, known as cohesion. Braze welding also uses a nonferrous filler metal, which melts at a lower temperature than the parent metal.

In both brazing and braze welding the parent metal is heated to a dull red heat. Then the filler rod is applied along with a suitable flux. Because the melting point of bronze is approximately 1600°F, it melts as it comes in contact with the hot parent metal.

Braze welding does not produce as strong a joint as fusion welding, because the bronze has less strength.   However it is a widely used process for many applications, such as joining cast iron.

Flux

Brazing flux, which has the appearance of melted glass, is usually made of a mixture of borax and boric acid. It is a cleaning agent which removes *mill scale* (an oxide coating, which forms on molten iron and steel) and other impurities from the surface of the metal. Flux can easily be cleaned from the weldment after it cools.

CAUTION: Bronze contains zinc, which gives off posionous fumes when it is heated. Inhaling these fumes causes stomach cramps, vomiting and shortness of breath. Do not breathe brazing fumes.

JOB 18: BRAZE WELDING SHEET METAL

Equipment and Materials

Standard oxyacetylene welding equipment
1 piece, 16-gauge mild steel 2" x 6"
1/16" or 3/32" Bronze filler rod
Brazing flux

PROCEDURE	KEY POINTS
1. Grind the mill scale from a metal sample, so that the entire surface is shiny.	1. Wear goggles while grinding.
2. Place the sample on firebrick.	
3. Adjust the torch to a slightly oxidizing flame.	3. A slightly oxidizing flame is normally used for brazing.
4. Heat about 2" of the end of a piece of bronze rod with the flame. Dip the heated rod into a can of flux, so that some of it sticks to the end of the rod.	4. Plenty of flux is the key to good brazing. **CAUTION:** Brazing fumes are poisonous. Avoid breathing them. Work in a well ventilated area and warn others, so that they can taken precautions.
5. Heat the right edge of the metal to a dull red, then add filler rod, melting it off in the heat of the torch, figure 18-1.	5. Keep plenty of flux on the filler rod, heating it and redipping it frequently.

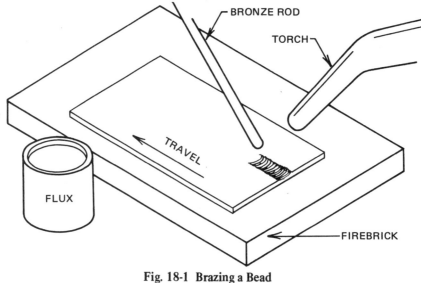

Fig. 18-1 Brazing a Bead
(Use Slightly Oxidizing Flame)

PROCEDURE	KEY POINTS
6. Keep the sample red hot and add bronze until the bead runs the full length of the plate.	6. Keep the flame small and point it in the direction of the weld. Do not melt the parent metal. Bronze will sputter and fume if overheated. Flash the torch off the bead to prevent overheating.
7. Practice until a bead can be applied with a smooth uniform appearance and with no evidence of burning.	7. The finished bead should have a shiny appearance and the metal should not look burned.

REVIEW QUESTIONS

1. Briefly describe what brazing is.
2. Why is flux used in brazing?
3. What is capillary attraction?
4. What is the difference between brazing and fusion welding?
5. How does the strength of a braze welded joint compare with that of a fusion welded joint?
6. Does bronze melt above or below the melting point of steel?
7. List the safety precautions which should be taken when braze welding.

Unit 19   Braze Welding Butt Joint on Mild Steel Sheet Metal

OBJECTIVES

The student will be able to:

- Gauge the proper heat for braze welding by using the parent metal color as an indicator.

- Properly clean and prepare metal for braze welding.

- Braze weld a butt joint capable of withstanding a bend test.

A braze welded joint is made in the same general fashion as a fusion welded joint. Braze welding requires special preparation which is not ordinarily used in the fusion process. The braze welded butt joint is common in many applications because it is faster than fusion welding and requires less training to complete. However, it is not as strong as a fusion welded joint.

Bronze rod will not flow properly if the metal is not thoroughly cleaned before brazing. Grinding the edges and both sides of the metal near the edge where the braze is to be made cleans the parent metal so that the bronze will flow through the joint and onto the bottom side.

Do not overheat the joint. Keep the color dull red and flash the torch away from the braze if it appears to be overheating. If the parent metal becomes molten, the bronze will burn into it and make the parent metal weak. Overheating the metal will also cause the bronze to drip through the joint and build up too much underbead.

> **CAUTION: The fumes given off by the brazing process are poisonous. No braze welding should be done unless proper ventilation is provided.**

JOB 19: BRAZE WELDED BUTT JOINT ON MILD STEEL SHEET METAL

Equipment and Materials

Standard oxyacetylene welding equipment
2 pieces, 16-gauge mild steel, 2″ x 6″
1/16″ or 3/32″ Bronze filler rod
Brazing flux

PROCEDURE	KEY POINTS
1. Clean the surfaces to be joined, by grinding.	1. Grind the edges to be butted and both the top and bottom surfaces.

PROCEDURE	KEY POINTS
2. Tack the two pieces in position for a butt joint, gapping the pieces uniformly about 1/16″, figure 19-1.	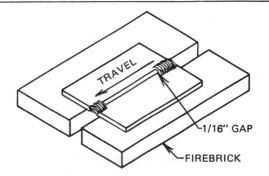

Fig. 19-1 Braze Welded Butt Joint

PROCEDURE	KEY POINTS
3. Apply flux to the rod and bring the parent metal to a red heat. Add rod to the metal, so that the bronze flows through the gap. Work the braze across the entire joint. 4. Test the braze in a vise, figure 19-2.	3. Do not overheat the metal. This is not a fusion weld; do not melt the parent metal. Build the bead up slightly above flush. 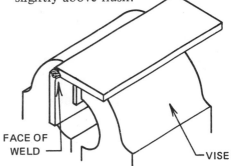

Fig. 19-2 Braze Welded Butt Joint Test
(Bend Against Face of Weld, Full 90°)

REVIEW QUESTIONS ·

1. Why is braze welding faster than fusion welding?

2. What special preparation of the metal is required before braze welding a butt joint?

3. Why is a braze welded joint not as strong as a fusion welded joint?

4. How is the temperature of the parent metal determined for braze welding?

5. Why is good ventilation necessary when brazing?

Unit 20   Braze Welding Cast Iron

OBJECTIVES

The student will be able to:

- List the properties of cast iron that make it more difficult to weld than other ferrous metals.

- Prepare the surfaces of cast iron for braze welding.

- Braze weld cast iron assemblies capable of withstanding destructive testing.

Cast iron is an alloy, containing from 1.7% to 4.5% carbon and small amounts of silicon.   The addition of large amounts of carbon to iron make it brittle and hard. Cast iron is easily formed into shapes, but being brittle, it will not withstand bending and twisting.   Because brazing is done at much lower temperatures than fusion welding, it causes less expansion and contraction. For this reason cast iron is often braze welded.

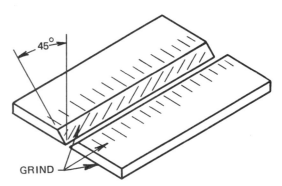

Fig. 20-1 V-Grooving and Grinding

V-Grinding

To present a wider surface for bonding and to assist in getting 100% penetration, the pieces are often *V-ground*. V-grinding refers to beveling the edges of the pieces to be joined, so that when they are put together the joint forms a V, figure 20-1.

The following tips will contribute to good, full-strength braze welded joints:

- Preheat the casting, so cracks do not appear as a result of expansion and contraction.

- Braze welding is not fusion welding. Do not melt the parent metal.

- 100% penetration of the bronze through the joint is required for full strength.

JOB 20: BRAZE WELDING CAST IRON

Equipment and Materials

Standard oxyacetylene welding equipment
2 pieces broken cast iron
3/32″ or 1/8″ Bronze filler rod
Brazing flux

PROCEDURE	KEY POINTS
1. Grind the edges of the two pieces of cast iron to a V of about 45°.	1. Grind the surfaces back from the edges of the V, so that the bead will be wider than the V.
2. Use a large tip, depending on the thickness of the cast iron.	2. Use a slightly oxidizing flame. If the puddle bubbles, there is too much acetylene in the flame.
3. Tack each end of the cast iron pieces so a small gap appears at the bottom of the V.	
4. Heat the area near the joint to a dull red.	4. Play the flame over the pieces. Be sure the entire casting is preheated.
5. Heat the bronze rod and dip it into the flux, then transfer the flux to the heated V.	5. If the joint is long, scatter flux along the bottom of the V by hand. The metal should be hot enough so the flux melts and is not blown off the joint by the torch.
6. While the parent metal is red-hot, melt a small amount of bronze into the V and allow it to flow over the surface of the clean metal.	6. If the bronze does not flow smoothly, the metal may not be hot enough. Applying a thin coat of metal in this manner is called *tinning*.
7. Continue to tin the area ahead of the bead, then back over the area, filling the joint with bronze.	
8. Do not work too rapidly. Cast iron must be at the right temperature to make the bronze bond with the cast iron.	8. Overheating is indicated by a white powder deposit and smoking. **CAUTION: Do not breathe zinc fumes.**
9. Cool the work slowly to avoid cracks.	
10. The bronze must completely penetrate the joint, forming a small bead on the bottom.	
11. Test the cooled assembly by breaking over the bead.	11. A good braze weld should break the parent metal outside the joint.

REVIEW QUESTIONS

1. What property of cast iron makes it more difficult to weld than other ferrous metals?

2. What characteristic of the braze welding process makes it better than fusion welding for joining cast iron?

3. Give two reasons why cast iron joints are often V-ground.

4. What is indicated if a white powder appears or if the braze smokes?

5. What is indicated if the bronze does not flow smoothly?

Unit 21 Backhand Welding on Mild Steel Sheet Metal

OBJECTIVES

The student will be able to:

- Explain how stresses develop in metal during welding.
- Describe backhand welding and torch angle.
- Discuss the use of backhand welding for stress relief.
- Weld a bead using the backhand method.

Backhand welding is frequently used for welding heavy plate and pipe joints. In backhand welding the flame is directed back against the completed weld. The filler rod is added into the flame at the head of the weld and the bead forms behind the filler rod. The direction of the weld is from left to right for right-handed welders (right to left for left-handed welders), figure 21-1.

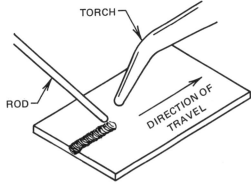

Fig. 21-1 Backhand Welding

Stress Relief

Heating and cooling can cause stresses to develop within the metal. Because the metal next to the weld cools at a different rate than the metal away from the weld, the expansion and contraction is not uniform. The resulting stresses can cause cracks to appear as the hotter metal is pulled away by the cooling process.

These stresses may be relieved by reheating the joint to a red heat and allowing it to cool slowly. In backhand welding the flame is directed onto the newly formed weld. This allows the weld to cool more slowly and stress relieves the joint.

JOB 21: BACKHAND WELDING ON MILD STEEL SHEET METAL

Equipment and Materials

Standard oxyacetylene welding equipment
16-gauge sheet metal sample, 2" x 6"
1/16" or 3/32" Mild steel filler rod

PROCEDURE	KEY POINTS
1. Start a molten puddle at the left end of the sample. (Left-handed welders start at the right end of the sample.)	

PROCEDURE	KEY POINTS
2. Add filler rod and continue moving the bead to the right.	2. Hold the torch at an angle of 30° to 45°, so the flame is directed back over the work which has been welded.
3. Run the bead across the entire 6" length of the sample.	3. Backhand welding requires practice.

REVIEW QUESTIONS

1. Describe the difference between forehand welding and backhand welding.

2. What causes stresses to develop in welding?

3. How does backhand welding help relieve stress?

4. Name two types of welding that are commonly done by the backhand method.

Unit 22 Backhand Butt Weld on Sheet Metal Mild Steel

OBJECTIVES

The student will be able to:

- List the advantages of backhand welding over forehand welding.
- Control penetration of a backhand weld by torch and filler rod manipulation.
- Make a backhand butt weld capable of withstanding bend testing.

Material for backhand butt welding is prepared in the same fashion as that for forehand welding. Two sheet metal samples are tacked together at each end, with approximately a 1/16" gap at one end and 1/8" gap at the other end. However, welding is done from left to right, so the sample should be set up for welding with the 1/16" gap at the left-hand side and welding should progress toward the 1/8" gap. The torch is pointed so that the flame points over the weld as it is being completed, figure 22-1.

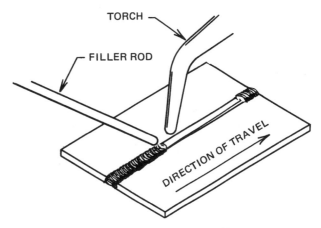

Fig. 22-1 Backhand Butt Weld

The filler rod should be added in front of the flame, and flashing off the weld may be necessary to control the amount of penetration. However, penetration of this weld, as of all oxyacetylene welds, must be 100%.

JOB 22: BACKHAND BUTT WELD ON MILD STEEL SHEET METAL

Equipment and Materials

Standard oxyacetylene welding equipment
2 pieces, 16-gauge mild steel, 2" x 6"
1/16" or 3/32" Mild steel filler rod

PROCEDURE	KEY POINTS
1. Tack weld the two samples of mild steel together for a butt weld.	1. Gap the pieces 1/16" at the left side and 1/8" at the right.

PROCEDURE	KEY POINTS
2. Beginning on the left side of the plates, weld them together using the backhand method.	2. Concentrate on smooth bead appearance and 100% penetration.
3. Test the weld by bending the plates 180° over the crown of the weld.	3. Penetration should be complete, with no cracks or holes.

REVIEW QUESTIONS

1. From what direction is the filler rod added in backhand welding?
2. Compare the appearance of a backhand bead with that of a forehand bead.
3. How much penetration is required in backhand welds?
4. What are the advantages of backhand welding over forehand welding?

SECTION 3: OVERHEAD WELDS, BRAZING AND BACKHAND WELDS, COMPREHENSIVE REVIEW

A. OVERHEAD BUTT WELD

Equipment and Materials

> Standard oxyacetylene welding equipment
> 2 pieces, 16-gauge mild steel, 2″ x 6″
> 1/16″ or 3/32″ Mild steel filler rod

PROCEDURE

1. Prepare the material for a standard butt weld. Leave a gap for contraction.
2. Clamp the material in an overhead position.
3. Weld the joint with one pass.
4. Test the weld by bending 180° over the bead.
5. Have the instructor inspect the weld.

B. BRAZE WELDED BUTT JOINT IN MILD STEEL

Equipment and Materials

> Standard oxyacetylene welding equipment
> 2 pieces, 16-gauge mild steel, 2″ x 6″
> 1/16″ or 3/32″ Bronze filler rod
> Brazing flux

PROCEDURE

1. Grind the edges to be joined and both faces near the edge.
2. Tack the two pieces in position for a butt joint.

3. Heat the filler rod and apply the flux to it.

4. Heat the parent metal to a red heat, then braze the joint.

5. Test the joint in a vise.

6. Have the instructor inspect the joint.

C. BACKHAND BUTT WELD

Equipment and Materials

Standard oxyacetylene welding equipment
2 pieces, 16-gauge mild steel, 2" x 6"
1/16" or 3/32" Mild steel filler rod

PROCEDURE

1. Prepare the material for a butt weld.

2. Position the material for backhand welding in the flat position.

3. Weld the joint backhand with one pass.

4. Test the weld by bending 180° over the bead.

5. Have the instructor inspect the weld.

SECTION: 4
Oxyacetylene Cutting

Unit 23   Straight Cutting with the Oxyacetylene Cutting Torch

OBJECTIVES

The student will be able to:

- Explain how an oxyacetylene flame cuts ferrous metal.

- Assemble an oxyacetylene cutting outfit.

- Make clean, smooth cuts in mild steel plate.

Oxyacetylene cutting is done by directing a stream of oxygen onto the ferrous metal, which has been preheated. The oxygen burns the metal. By controlling the amount of pre-heat and the size of the stream of oxygen, a cut may be made with clean, smooth sides. The cut is called a *kerf*

Oxyacetylene cutting is one of the most used oxyacetylene processes. The cutting torch can be used to cut intricate shapes or to make straight, clean cuts. The cutting, or burning, process does not change the chemical composition of the metal. Therefore, any ferrous metal can be welded immediately after it has been cut. However, *slag* (oxidized metal) is sometimes left at the bottom edge of the cut. This must be removed by grinding or chiseling. If oxidized metal is included in the puddle it will contaminate the weld. Metal should always be clean before welding.

Cutting Torch

The cutting torch, figure 23-1, is designed only for cutting ferrous metals. To install the cutting torch the welding torch handle must be removed from the hoses. The cutting torch is then installed in its place. The cutting torch is designed for heavy cutting and per-forms better over long periods of time than the cutting head.

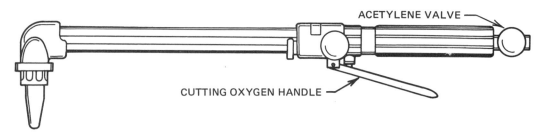

Fig. 23-1 Cutting Torch

Fig. 23-2 Cutting Head   Fig. 23-3 Cutting Tip   Fig. 23-4 Four-Hole Cutting Tip, Installed for Straight-Line Cut. Make Cut Right to Left.

Cutting Head

The cutting head, figure 23-2, is an attachment to the welding torch. By removing the tip from the welding torch the cutting head may be screwed onto the torch handle. In this manner a welding torch may be used for cutting.

Oxygen Pressure

Since large amounts of oxygen are required to burn the metal, more oxygen pressure is needed for cutting than for welding. When using a cutting torch, or cutting head, the gas pressures should be regulated according to the manufacturer's specfications for the torch being used.

Cutting Tip

The cutting tip, figure 23-3, is designed especially for cutting and cannot be used for welding. Cutting tips are made with a hole in the center, through which the stream of oxygen is directed at the cut. A group of holes around the center hole give off a neutral flame which preheats the metal, figure 23-4. Depending on the size of the tip, there are 4, 6, 8, or 12 preheat holes. Each of these is like a miniature welding tip and when the torch is lighted, it should be adjusted so that each of the preheat holes makes a neutral flame.

When the metal to be cut has been preheated to red-hot, the cutting oxygen valve is pressed. The stream of oxygen will burn (cut) the metal as long as the preheat is maintained. Oxyacetylene cutting must be done at a slow, even rate of speed. If the cut is made too rapidly, the metal may cool down and the cutting action will stop. If this happens the torch should be moved back into the kerf and the metal preheated again. The cut may then be started again.

JOB 23: OXYACETYLENE CUTTING

Equipment and Materials

Oxyacetylene outfit with cutting head
Straightedge
Hammer
Center punch
Soapstone or chalk
Scrap pieces of 1/2-inch mild steel plate

PROCEDURE	KEY POINTS
1. Install the cutting head on the torch handle.	1. The procedure for using a cutting torch is the same as for a cutting head.
2. Install the cutting tip in the position shown in figure 23-4.	
3. Use the acetylene and oxygen pressures recommended by the manufacturer of the equipment.	3. The high pressure of the cutting oxygen helps blow the kerf clear.
4. Light the cutting torch in the same manner as the welding torch and adjust it for a neutral flame.	4. Wear gloves and goggles. All preheat flames should be neutral.
5. Depress the cutting oxygen valve and with the valve wide open, adjust the preheat flames to neutral.	5. With the cutting oxygen valve wide open the working pressure should be adjusted to the manufacturer's specifications.
6. Shut off the torch.	
7. Mark a line 1/2 inch from the end of the steel plate, using the straightedge and soapstone.	7. Center punch the line. Soapstone or chalk will burn off, but center punch marks can be followed after the soapstone is gone.
8. Position the plate on a suitable table so that the marked end overhangs the edge of the table.	
9. Relight the cutting torch.	
10. Hold the flame about 1/8" above the edge of the plate.	10. Position the flame at the beginning of the line, but do not let it touch the plate.
11. Preheat the metal to a red heat.	11. Brace the torch by using the bench behind the plate for support.
12. Depress the oxygen trigger and, as the metal burns and is blown through the kerf, move across the plate. Follow the marked line until the cut is completed.	12. **CAUTION: Stand clear of the material; when the cut is completed the material will fall off. Falling sparks can ignite clothing. Do not cut or direct the stream of oxidized material toward inflammable objects or toward oxygen and acetylene containers.**
13. Practice cutting pieces from the plate until a smooth, even, straight cut is achieved.	

Summary: Job 23

- The preheat flames must be kept neutral.
- The torch must be held at an angle of 90° with the cut so that a straight edge is made.
- Keep the oxygen trigger fully depressed so that the kerf blows clean.
- The tip must be kept clean. Use tip cleaners if the holes become plugged.
- Be sure the tip is installed for straight cutting.
- The metal must be kept red-hot, or the cutting action will stop.

REVIEW QUESTIONS

1. How does the oxyacetylene flame cut metal?
2. How is a cutting head installed on an oxyacetylene welding outfit?
3. What is the result when a cut is made too fast and the preheat is lost?
4. What is a kerf?
5. What is slag?
6. Make a sketch of the end of the cutting nozzle, showing how it is installed for a straight cut.

7. Is the acetylene pressure increased when the cutting torch is used?
8. Why should a line be center-punched for cutting with the cutting torch?
9. List the safety measures which should be taken when cutting with the torch?

Unit 24   Beveling Plate with the Oxyacetylene Cutting Torch

OBJECTIVES

The student will be able to:

- Describe the position of the cutting torch tip for cutting a bevel.
- Define a land and tell why it is used.
- Cut a straight, smooth bevel with the oxyacetylene torch.

Ordinarily, 100% penetration is required of an oxyacetylene weld in order to insure strength in the welded joint. Metal which is over 1/8 inch thick is very difficult to melt through, so some method has to be provided to insure complete penetration. Metal 1/8 inch to 3/16 inch thick is frequently gapped for welding, but the edge of metal over 3/16 inch thick should be *beveled*. This is done by cutting the edge of the metal on an angle.

Cutting straight through a piece of steel leaves a cross section the same width as the thickness of the original metal. However, when the edge is beveled, the cross section is increased, figure 24-1.

Beveling leaves the bottom of the plate with a very thin edge, which has a tendency to melt off during welding. This edge is generally ground square to a thickness of 1/16 inch, to prevent the edge from melting off. This ground shoulder is called a *land*.

Sometimes when bevels are made, a small increase of oxygen pressure is necessary to cut the larger cross section of the bevel. To cut a bevel, the cutting tip should be turned so the holes line up as shown in figure 24-2.

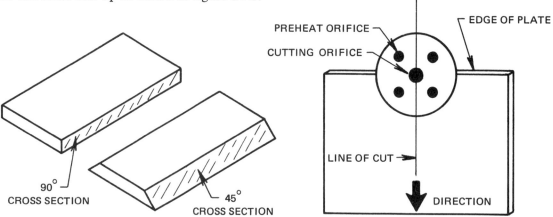

Fig. 24-1 Cross Section of Straight and Beveled Edges

Fig. 24-2 Cutting Tip Installed for Bevel Cut

JOB 24: BEVELING PLATE WITH THE OXYACETYLENE CUTTING TORCH

Equipment and Materials:

Standard oxyacetylene outfit with cutting head
Straightedge
Hammer
Center punch
Soapstone
Scrap pieces of 1/2" thick mild steel plate

PROCEDURE	KEY POINTS
1. Mark a straight line with soapstone and a center punch.	
2. Preheat the edge of the plate and cut along the mark, with the torch held at a 45° angle with the plate.	2. Be sure the cutting tip is installed on the torch correctly for a bevel cut. Keep the torch the same distance from the plate at all times. Move slowly and steadily, so the preheat is not lost. Visibility will be best if the torch is drawn toward the operator.
3. Practice cutting a bevel until a smooth, regular bevel is achieved.	3. An angle iron guide may be used to maintain a straight, even bevel, figure 24-3.

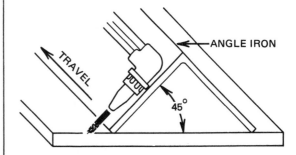

Fig. 24-3 Method of Using Angle Iron to Cut Bevel

REVIEW QUESTIONS

1. Why is metal over 3/16 inch thick beveled to prepare it for a butt weld?
2. How does beveling affect the cross section of a piece of metal?
3. What is the land of a beveled plate?
4. If a piece of angle iron is used as a guide for cutting a bevel what will the angle of the bevel be?
5. Draw a sketch showing the placement of the preheat holes in a cutting tip as it is used for beveling metal.

Unit 25   Cutting Holes with the Oxyacetylene Cutting Torch

OBJECTIVES

The student will be able to:

- Pierce steel plate with the oxyacetylene cutting torch.
- Cut round holes within 1/16 inch of the right diameter.
- Holes cut will have clean, straight sides and be free of slag.

The oxyacetylene cutting torch is a good tool for cutting holes in steel, where a precision fit is not necessary. Round, square, rectangular, and odd-shaped holes can be cut equally well.

As the cutting oxygen valve is opened, after the metal has been preheated to a red heat, the torch tip must be

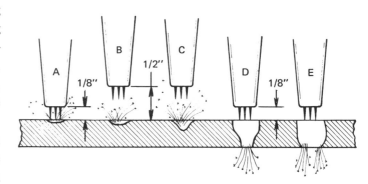

Fig. 25-1 Torch Handling Sequence for Hole Piercing

moved upward away from the cut. This is to keep the slag from blowing up into the tip. When the hole is burned completely through the plate, lower the torch until the preheat flames are about 1/8 inch from the surface of the metal, figure 25-1. With the hole pierced and the torch in position to keep the metal preheated, cut away from the hole to the mark and around the circle. Clean the slag from the underside when the cut is completed.

JOB 25: CUTTING HOLES WITH THE OXYACETYLENE CUTTING TORCH

Equipment and Materials

> Standard oxyacetylene cutting outfit
> Soapstone
> Center punch
> Hammer
> Scraps of 1/4-inch or 1/2-inch mild steel plate

PROCEDURE	KEY POINTS
1. Mark 1-inch and 2-inch circles on the plate, using soapstone.	1. Center punch the soapstone lines.
2. With a neutral flame, preheat the inside of the circle. When the metal is red-hot, depress the oxygen lever.	

PROCEDURE	KEY POINTS
3. When the cut starts, raise the tip up about 1/2″.	3. This prevents the oxide from blowing back into the tip.
4. When the hole has been burned through, lower the torch until the preheat flame is about 1/8″ from the surface of the metal.	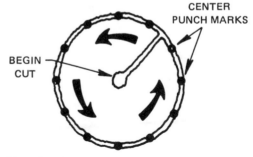
5. Cut outward to the punched line and follow the mark around the circle, figure 25-2.	
6. Practice this until a smooth, slag-free cut is achieved.	**Fig. 25-2 Cut from Center Outward to Rim**

REVIEW QUESTIONS

1. What type of preheat flame is used for oxyacetylene cutting?

2. What is done to prevent the slag from blowing up into the tip as a hole is started?

3. When cutting a hole in the center of a plate, where is the cut started?

4. When making a cut, how far should the preheat flames be from the surface of the metal?

5. How can the operator tell when the metal is preheated enough to begin a cut?

SECTION 4: OXYACETYLENE CUTTING, COMPREHENSIVE REVIEW

A.   OXYACETYLENE FLAME CUTTING

Equipment and Materials

Standard oxyacetylene cutting outfit
Soapstone
Center punch
Hammer
1/4-inch mild steel plate

PROCEDURE

1. Mark all cuts with soapstone and center punch the marks. See oxyacetylene cutting evaluation drawing.

2. Complete all cuts slowly and carefully.

3. Remove all slag.

4. If the cuts are not acceptable, practice cutting on scrap until enough skill is developed to complete this evaluation.

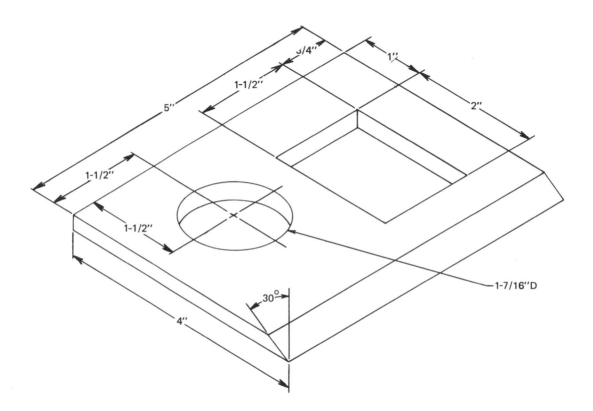

SECTION 5:
Welding Steel Plate and Pipe

Unit 26   Welding Mild Steel Plate

OBJECTIVES

The student will be able to:

- Explain the disadvantage of using oxyacetylene for welding heavy plate.

- Prepare steel plate for a butt weld.

- Make a butt weld having good appearance and penetration in steel plate.

Oxyacetylene welding is not generally used to join plates over 3/16 inch thick. It is, in fact, most often used for metals less than 1/8 inch thick. However, the method described in this unit may be used for any thickness of metal.

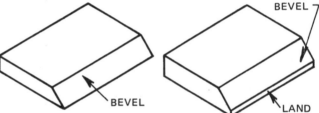

Fig. 26-1 Bevel 30° to 45°   Fig. 26-2 Land Ground on Bevel

Plate (metal over 1/8 inch thick) welding requires concentration of the heat. For this reason, most plate welding is done by the electric arc process.

Preparation

It is impossible to achieve 100% penetration of thick metal without leaving a gap between the pieces, or grinding the edges. For this reason, plate preparation is as important as the weld itself. Steel plate is beveled, to insure 100% penetration. This is done by grinding the edges at an angle, figure 26-1.

Beveling leaves a very thin edge, which can melt off during welding and allow excessive penetration. To prevent this edge from melting off, a land, or square shoulder, is ground on the bevel, figure 26-2.

JOB 26A: WELDING MILD STEEL PLATE

Equipment and Materials

Standard oxyacetylene welding equipment
2 Mild steel plates, 3/16″ x 4″ x 6″
1/8-inch mild steel filler rod

PROCEDURE	KEY POINTS
1. Grind a 45° bevel on one edge of each plate.	1. A bevel is necessary for 100% penetration.
2. Place the material on a firebrick so that the sides are separated about 1/8″ at the left end and 1/16″ at the right end, figure 26-3. (Left-handed welders reverse this position.)	2. Gapping the plates allows for expansion and contraction caused by the welding heat.

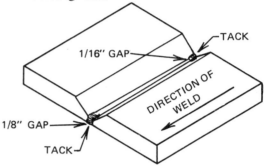

Fig. 26-3 Setup for Butt Weld

3. Tack the plates securely at each end.	3. Tack welds should have 100% penetration and good fusion.
4. Place the heat over the right end of the joint until the metal is molten. Hold the end of the filler rod in the cone of the flame to heat it for welding.	
5. When the molten metal penetrates the gap entirely, add the filler rod, moving the torch in a slow arc.	5. Insure 100% penetration at all times, but do not overheat the metal enough to cause excessive drop-through. It is important that fusion is good at the bottom edge of the plates.
6. Continue in this manner until the weld is completed.	6. Stir the molten metal by working the torch in a circular motion, keeping the metal rolling. Thoroughly mix the metal from both plates to insure good fusion.
7. Allow the sample to cool slowly.	

JOB 26B: TESTING A BUTT WELD IN MILD STEEL PLATE

Equipment and Material

Oxyacetylene cutting equipment or metal saw
Straightedge
Soapstone
Center punch
Hammer

PROCEDURE	KEY POINTS
1. With the soapstone and straightedge, mark three one-inch strips the length of the plate, figure 26-4.	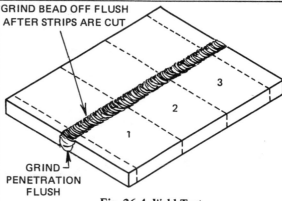 Fig. 26-4 Weld Test
2. Carefully cut the strips from the plate with the cutting torch or the saw.	2. If the cutting torch is used, be careful not to leave deep gouges in the edge of the strips. Discard the outside strips.
3. Grind the slag and crown of the bead off from the plate, so that it is the original thickness for its entire length.	3. All grinding marks should be parallel to the length of the plate. Gouges or grinding marks across the plate will cause it to fail at that point.
4. Place the center strip in a tensile tester and pull until the sample breaks.	4. The weld should hold more than the parent metal.
5. Bend one of the outside strips against the penetration and the other with the penetration.	5. Both samples should bend 180° without failure.

REVIEW QUESTIONS

1. What is the disadvantage of oxyacetylene for welding heavy steel plate?

2. Why must the edges of plate be specially prepared for welding?

3. What is the land of a beveled plate?

4. What is plate?

5. What is a bevel?

6. What type of flame is used for plate welding?

7. Why should a circular motion of the torch be used to weld plate?

8. How much penetration is required for a good plate weld?

Unit 27   Butt Welding Pipe in the Vertical Fixed Position

OBJECTIVES

The student will be able to:

- Describe the vertical fixed position for pipe welding.
- Properly prepare pipe for welding in the vertical fixed position.
- Butt weld pipe in the vertical fixed position, making a joint capable of withstanding a 180° bend.

The oxyacetylene welder should be able to weld mild steel pipe of 2 inch or less diameter. Most welding on pipe which is over 2 inch diameter is done with electric arc welding, because of speed of production. However, a great amount of 1-, 1-1/2 and 2-inch pipe is welded with the oxyacetylene torch.

It is sometimes possible to make pipe butt joints where the pipe can be rolled to keep the weld on the topside and easily available. However, many times the pipe cannot be moved, and the welder must work around the pipe, changing the angle of the torch to fit the pipe circumference. This is called

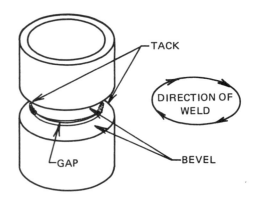

Fig. 27-1 Butt Weld, Pipe in the Vertical Fixed Position

fixed position welding. When a pipe is within 10 degrees of vertical and in the fixed position, it is in the vertical fixed position, figure 27-1.

The pipe ends should be grooved to insure 100% penetration of the butt joint. Also, the pieces must be solidly tacked together, to hold them in a straight line.   With the pipe in the vertical position, the weld will be horizontal.

JOB 27: BUTT WELDING PIPE IN THE VERTICAL FIXED POSITION

Equipment and Materials

Standard oxyacetylene welding equipment
Jig for holding pipe sample
Oxyacetylene cutting torch or metal saw.
1-inch pipe samples, 8″ long
1/8-inch mild steel filler rod

PROCEDURE	KEY POINTS
1. Grind the ends of the two pipe samples to a 30° bevel.	

PROCEDURE	KEY POINTS
2. Tack weld the beveled ends together.	2. Keep pipe samples in a straight line when tacking. Tacks should have 100% penetration and good fusion.
3. Position the pipe in the vertical fixed position.	3. Welding pipe in the vertical fixed position is the same as making a horizontal butt weld, except that the weld travels around the pipe, instead of across the plate.
4. Weld around the pipe, walking around it.	4. Keep a keyhole at all times. Overlap the beginning and end of the weld.
5. Cut three 1/2-inch strips from the welded pipe and bend test them in a vise, figure 27-2.	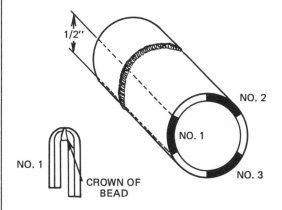 Fig. 27-2 Pipe Weld Test
6. If the strips break in the bend test, examine the weld and determine the reason for failure. **CAUTION: Be careful with the lighted torch when walking around the joint. Be especially careful of other people in the area when flashing off the weld.**	6. Incomplete penetration or too much heat may cause failure. Pipe joints with incomplete penetration may leak under pressure.

REVIEW QUESTIONS

1. Why are the ends of pipe grooved in preparation for welding?

2. In the vertical fixed position does the weld run in a vertical or horizontal direction?

3. Why are lengths of pipe tacked together before they are welded?

4. Other than making a weak joint, what might be the result of poor penetration in a pipe weld?

5. What sizes of pipe are most often welded with the oxyacetylene torch?

Unit 28   Butt Welding Pipe in the Horizontal Fixed Position

OBJECTIVES

The student will be able to:

- Describe the horizontal fixed position for pipe welding.
- Overlap beads on a pipe weld with complete fusion.
- Weld pipe in the horizontal fixed position, making a joint capable of withstanding a 180° bend.

When pipe is welded in the horizontal position it is often impossible to roll the pipe for welding. The welder must make an acceptable joint with the pipe in the horizontal fixed position. This is the position when the pipe is within 30° of horizontal and cannot be rolled.

To achieve complete penetration, begin the weld at the bottom of the joint and weld vertically around the side to the center of the top. Then begin at the bottom and weld up the other side to complete the weld, figure 28-1. Keep a keyhole in front of the weld for penetration.

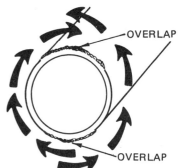

Fig. 28-1 Welding Pipe in the Horizontal Fixed Position

Whenever a weld overlaps another, great care must be taken to thoroughly melt the beads together for good fusion. Incomplete fusion (cold laps) will cause weld failure or leakage.

JOB 28: BUTT WELDING PIPE IN THE HORIZONTAL FIXED POSITION

Equipment and Materials

Standard oxyacetylene welding equipment
Jig for holding pipe sample
Oxyacetylene cutting torch or metal saw
1-inch pipe samples, 8″ long
1/8-inch mild steel filler rod

PROCEDURE	KEY POINTS
1. Bevel one end of the two pipes.	
2. Tack the two beveled sections together.	2. Make sure samples are in a straight line.
3. Position the sample in the horizontal fixed position, figure 28-2.	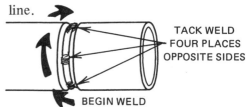 Fig. 28-2 Horizontal Fixed Pipe Weld

PROCEDURE	KEY POINTS
4. Start the weld at the bottom of the groove, on the under side.	4. Most of the weld is vertical. Make sure the overlaps are completely fused. Keep a keyhole for 100% penetration.
5. Cut three 1/2 inch wide strips from the pipe and bend test them in the vise.	
6. If the strips break in the bend test, examine them and determine the cause of failure.	

REVIEW QUESTIONS

1. What is meant by beveling?
2. What is meant by fixed position?
3. Is a horizontal pipe butt weld made from top to bottom, or bottom to top?
4. What precaution must be taken when welding an overlap on a bead?
5. How much penetration is required for pipe butt joints?
6. Why is a keyhole kept open on a butt weld?

SECTION 5: WELDING MILD STEEL PLATE AND PIPE, COMPREHENSIVE REVIEW

A. WELDING MILD STEEL PLATE

Equipment and Materials

Standard oxyacetylene welding equipment
Two samples, mild steel plate 1/4" x 4" x 6"
1/8-inch mild steel filler rod

PROCEDURE

1. Bevel one edge of each plate.
2. Tack weld the plates for a butt joint.
3. Play the heat on the end of the joint until a puddle is established.
4. Add filler rod and move the torch in a slow arc, being sure to achieve 100% penetration.
5. Continue welding in this manner until a slight crown is built up.
6. When the weld has cooled, grind both sides flush with the surface of the parent metal.
7. Cut three test strips from the weldment.
8. Perform a tensile test on one strip and bend the other two in opposite directions.

B. WELDING 2-INCH PIPE

Equipment and Materials

> Standard oxyacetylene welding equipment
> Jig for holding samples
> Oxyacetylene cutting torch or metal saw
> Four samples of 2-inch pipe, 8 inches long
> 1/8-inch mild steel filler rod

PROCEDURE

1. Bevel one end of each piece of pipe, and tack weld for two assemblies.
2. Position one sample for a vertical fixed pipe weld and the other for a horizontal fixed pipe weld.
3. Weld both samples.
4. Cut three strips from each sample.
5. Bend test the strips and have them checked by the instructor.

Glossary

Backhand Welding: A method of welding in which the torch is directed in the opposite direction from that of the welding progress.

Bead: The line of metal deposited by welding.

Bevel: To grind the edges of a plate at an angle for welding.

Brazing: A process using a filler metal which melts at a lower temperature than the parent metal, but above 800 degrees Fahrenheit. Brazing relies on capillary attraction instead of fusion.

Bridging: A condition resulting when the parent metal is not completely fused into the corner of a fillet weld, and the bead has been built across the gap.

Butt Weld: A weld joining the edges of two pieces of metal which are in line with one another.

Calcium Carbide: A gray, rock-like substance obtained by smelting coke and lime in an electric furnace. It produces acetylene gas when mixed with water.

Carburizing Flame: An oxyacetylene flame containing an excess of acetylene. A carburizing flame is characterized by an acetylene feather.

Cold Lap: Incomplete fusion between the filler rod and the parent metal.

Crown: The buildup of the bead above the thickness of the parent metal.

Cutting Head: An attachment to be used on a welding torch. The cutting head has a cutting oxygen valve and holds cutting tips, so that it can be used as a cutting torch.

Cutting Tip: A special tip to be used on a cutting torch or a cutting head, for cutting ferrous metals. The cutting tip has a center, cutting oxygen hole, and several preheat holes.

Cutting Torch: A specially designed torch to be used for flame cutting ferrous metals.

Cylinder Valve: A valve installed in the top of acetylene and oxygen cylinders. The cylinder valve is used to turn on and off the flow of gas from the cylinder.

Ductile Strength: The ability of a metal to withstand breakage.

Filler Rod: Also called welding rod. It is generally of the same metallic composition as the parent metal.

Fillet Weld: An inside corner weld made where two pieces of metal join at any angle. When one plate forms a 90 degree angle with the other, it is often called a T-weld.

Flash Off: Momentarily removing the flame from a molten puddle when the melting process becomes too rapid.

Flat Position: The position of a weld made on the topside of the parent metal and within 30 degrees of horizontal.

Flux: A cleaning agent which removes impurities from the surface of metal for brazing.

Forehand Welding: A method of welding in which the torch is directed in the direction that the weld progresses.

Fusion Welding: The complete mixing of two or more pieces of metal, melted to a liquid state by the oxyacetylene torch, and allowed to cool.

Goggles: Eye protection devices made from spark-resistant material, equipped with shaded filter lenses.

Horizontal Position: A weld made within 45 degrees of horizontal and against a vertical surface.

Jig: Any device used for holding material while work is being done. (A holding fixture).

Kerf: The cut made by any tool, such as a saw or cutting torch.

Keyhole: A small hole remaining open ahead of a butt weld to insure penetration.

Land: A small square edge ground at 90 degrees with the bottom edge of a bevel. It is also referred to as the "nose" of a bevel.

Lap Weld: A weld made on two pieces of metal with the edge of one overlapping the other.

Melt Weld: A weld made by fusing the parent metal without adding filler rod.

Mild Steel: Steel which contains 0.30% carbon, or less.

Mill Scale: Refers to the surface coat of oxide, which forms on molten iron or steel. Mill scale turns black when cool, giving metal the black coating which leads to the term "black iron".

Neutral Flame: A flame resulting from a balance of oxygen and acetylene. A neutral flame is used for all oxyacetylene fusion welding.

Non-Ferrous Metal: Any metal which does not contain iron. (Aluminum, copper, brass, etc.)

Overhead Position: The position of a weld made on the underside of the parent metal and within 45 degrees of horizontal.

Oxidizing Flame: An oxyacetylene flame containing an excess of oxygen. An oxidizing flame is characterized by a short inner cone and a whistling sound.

Parent Metal: The metal being welded.

Penetration: The depth the welding puddle melts into the parent metal.

Plate Metal: Metal which is more than 1/8 inch in thickness.

Regulator: A device which reduces the pressure of gas coming from a cylinder to the desired working pressure. Separate regulators are required for acetylene and oxygen.

Sheet Metal: Metal which has been rolled in the steel mills to a thickness of 1/8 inch, or less.

Slag: Oxides which are usually deposited on the surface of the metal after flame cutting or welding.

Striker: A tool used to light the torch. It makes a spark by moving a piece of flint across a file.

Tack Welding: To hold two or more pieces of metal in position for welding, small, completely fused welds (tacks) are made at various places along the line of the weld.

Tensile Strength: The ability to withstand a pull.

Torch: A welding tool with inlets for oxygen and acetylene, valves to control the flow of the gases, and a means of attaching the tip.

Torch Angle: The angle at which the torch tip and flame are pointed into the weld.

Undercut: The thinned, metal section resulting when fluid metal drops from the vertical surface of a weld.

V-Grinding: Grinding the edges of pieces to be welded so that a V is formed when they are put together.

Vertical Position: The position of a weld which is made on a vertical surface with the bead perpendicular to the ground or floor.

Welding Tip: A torch attachment equipped with a mixing chamber and orifice to regulate the size of the flame. It concentrates the gases coming from the torch so the flame can be directed toward the weld.

Weldment: An assembly of two or more parts which have been welded together.